CALCULUS:
A complete introduction

P. Abbott & Hugh Neill

First published in Great Britain in 1992 by Hodder Education. An Hachette UK company.

First published in US in 1992 by The McGraw-Hill Companies, Inc.

This edition published 2013 by Hodder & Stoughton. An Hachette UK company.

Previously published as *Teach Yourself Calculus* and *Understand Calculus: Teach Yourself*.

British Library Cataloguing in Publication Data: a catalogue record for this title is available from the British Library.

Library of Congress Catalog Card Number: on file.

10 9 8 7 6 5 4 3 2 1

The publisher has used its best endeavours to ensure that any website addresses referred to in this book are correct and active at the time of going to press. However, the publisher and the author have no responsibility for the websites and can make no guarantee that a site will remain live or that the content will remain relevant, decent or appropriate.

The publisher has made every effort to mark as such all words which it believes to be trademarks. The publisher should also like to make it clear that the presence of a word in the book, whether marked or unmarked, in no way affects its legal status as a trademark.

Every reasonable effort has been made by the publisher to trace the copyright holders of material in this book. Any errors or omissions should be notified in writing to the publisher, who will endeavour to rectify the situation for any reprints and future editions.

Cover image © Christian Delbert - Fotolia

Typeset by Cenveo® Publisher Services.

Printed in Great Britain by CPI Group (UK) Ltd., Croydon, CR0 4YY.

Hodder & Stoughton policy is to use papers that are natural, renewable and recyclable products and made from wood grown in sustainable forests. The logging and manufacturing processes are expected to conform to the environmental regulations of the country of origin.

Hodder & Stoughton Ltd
338 Euston Road
London NW1 3BH
www.hodder.co.uk

Contents

Introduction

Calculus is sometimes defined as 'the mathematics of change'. The whole point of calculus is to take the conventional rules and principles of mathematics and apply them to dynamic situations where one or more variable is changing. In particular, calculus provides you with the tools to deal with *rates* of change.

There are actually *two* 'big ideas' in calculus, called differentiation and integration. These are both techniques that can be applied to algebraic expressions. As you will discover, one is the converse of the other. Algebraically, what this means is that, if you start with some function $f(x)$ and differentiate it, then integrating the result will get you back, roughly speaking, to your original function $f(x)$. And the same idea is true in reverse – if you first choose to integrate $f(x)$, then differentiating the result will get you back to your original function $f(x)$. The notation used for integration is the sign $\int$, which is the old-fashioned, elongated letter 'S'.

But why might you want to perform a differentiation or an integration? Well, differentiating a function $f(x)$ gives a new function that tells you about the rate of change of $f(x)$. Because rate of change is a way of calculating the gradient of the graph of a function, this information allows you to explore many aspects of the function's shape. Integrating the function $f(x)$ allows you to find the area contained between the graph of $f(x)$ and the x-axis. A useful extension is that by imagining rotating, in three dimensions, the graph of a function around the x-axis (or the y-axis), integration can allow you to calculate the volume of the 'solid of revolution' so formed. This is a useful tool for finding the volumes of awkward shapes.

1

Functions

In this chapter you will learn:

▶ *what a function is, and how to use the associated notation*

▶ *how to draw the graph of a function*

▶ *how to use the notation δx to mean an increase in x.*

1.1 What is calculus?

Calculus is the Latin word for 'stone', and stones were used by the Romans for calculations. Here is a typical problem of calculus.

Consider a small plant which grows gradually and continuously. If you examine it after an interval of a few days, the growth will be obvious and you can measure it. But if you examine it after an interval of a few minutes, although growth has taken place, the amount is too small to see. If you observe it after an even smaller interval of time, say a few seconds, although you can detect no change, you know that the plant has grown by a tiny amount.

You can see the process of gradual and continuous growth in other situations. What is of real importance in most cases is not the actual amount of growth or increase, but the **rate of growth** or **rate of increase**. The problem of finding the rate of growth is the basis of the **differential calculus**.

Historical note: Sir Isaac Newton, in England, and Gottfried Leibnitz, in Germany, both claim to have invented the calculus. Leibnitz published an account in 1684, though his notebooks show that he used it in 1675; Newton published his book on the subject in 1693, but he told his friends about the calculus in 1669. It is generally accepted now that they discovered it independently.

1.2 Functions

You will know, from algebra, that the example in Section 1.1 of the growth of a plant is a **function**. The **growth** is a **function of time,** but you may not be able to give an equation for the function.

As the idea of a function is fundamental to calculus, this section is a brief revision of the main ideas.

When letters are used to represent quantities or numbers in an algebraic formula, some represent quantities which may take many values, while others remain constant.

Example 1.2.1

In the formula for the volume of a sphere, namely, $V = \frac{4}{3}\pi r^3$, where V is the volume and r is the radius of the sphere:

▶ V and r are different for different spheres, and are often called variables

▶ π and $\frac{4}{3}$ are constants, and do not depend on the size of the sphere.

Nugget – Spotting variables

In general, when you look at a formula, it is clear that the variables are the letters (conventionally written in italics) and the constants are the numbers. However, just to keep you on your toes, there are some constants that appear to masquerade as variables. One of these is pi, written π. Remember that pi is a constant; its value never changes. Another example of a constant disguised as a letter is the acceleration due to gravity, g. On the Earth's surface, the standard value of gravitational acceleration (at sea level) is 9.80665 metres per second per second, or approximately 9.81 ms^{-2}.

Example 1.2.2

In the formula for a falling body, namely, $s = \frac{1}{2}gt^2$, in which s is the distance fallen in time t, the letters s and t are **variables**, while $\frac{1}{2}$ and g are **constants**.

In each of the above examples, there are two kinds of variables.

In $V = \frac{4}{3}\pi r^3$, if the radius r is increased or decreased, the volume V increases or decreases as a result. That is, the volume V depends on r.

Similarly, in $s = \frac{1}{2}gt^2$, the distance s fallen depends on the time t.

Generally, in mathematical formulae, there are two types of variable, dependent and independent.

▶ the variable whose value depends upon the value of the other variable, is called the **dependent variable**, as in the case of V and s above

- the variable in which changes of value produce corresponding changes in the other variable is called the **independent variable**, as r and t above.

In the general quadratic expression

$$y = ax^2 + bx + c$$

where a, b and c are constants, as the value of y depends on the value of x, the independent variable is x and y is the dependent variable.

When two variables behave in this way, that is, the value of one variable depends on the value of the other, the dependent variable is said to be a function of the independent variable. In the examples above:

- the volume of a sphere is a function of its radius;
- the distance fallen by a body is a function of time.

Example 1.2.3

Here are some examples of the use of the word 'function'.

The square of a number is a *function* of the number.

The volume of a mass of gas is a *function* of the temperature while the pressure remains constant.

The sines and cosines of angles are *functions* of the angles.

The time of beat of a pendulum is a *function* of its length.

Definition: If two variables x and y are related in such a way that, when any value is given to x, there is one, and only one, corresponding value of y, then y is a **function** of x.

1.3 Equations of functions

When writing functions, letters such as x and y are usually used as the variables. For example, if $y = x^2 + 2x$, then, when a value is given to x, there is always a corresponding value of y, so y is a function of x. Other examples are:

$$y = \sqrt{x^3 - 5} \text{ for } x \geq \sqrt[3]{5}$$
$$y = \sin x + \cos x$$
$$y = 2^x$$

It is usual when working with functions to use letters at the end of the alphabet for the variables and letters towards the beginning or the middle of the alphabet for constants. Thus, in the general form of the equation of the straight line

$$y = mx + c$$

x and y are variables, and m and c are constants.

When expressing functions of angles, in addition to x, the Greek letters θ (theta) and ϕ (phi) are often used for angles.

1.4 General notation for functions

When it is necessary to denote a function of x without specifying its precise form, the notation $f(x)$ is used. In this notation, the letter f is used to mean the function, while the letter x is the independent variable. Thus $f(\theta)$ is a general method of indicating a function of θ.

Similar forms of this notation are $F(x)$ and $\phi(x)$.

A statement such as

$$f(x) = x^2 - 7x + 8$$

defines the specific function of the variable concerned. You would use this notation when you want to substitute a value of x.

For example, if $f(x) = x^2 - 7x + 8$, $f(1)$ means the numerical value of the function when 1 is substituted for x.

Thus
$$f(1) = 1^2 - 7 \times 1 + 8 = 2$$
$$f(2) = 2^2 - 7 \times 2 + 8 = -2$$
$$f(0) = 0^2 - 7 \times 0 + 8 = 8$$
$$f(a) = a^2 - 7a + 8$$
$$f(a + h) = (a + h)^2 - 7(a + h) + 8$$

Example 1.4.1

Suppose that $\phi(\theta) = 2\sin\theta$. Then

$$\phi\left(\tfrac{1}{2}\pi\right) = 2\sin\left(\tfrac{1}{2}\pi\right) = 2$$

$$\phi(0) = 2\sin 0 = 0$$

$$\phi\left(\tfrac{1}{3}\pi\right) = 2\sin\left(\tfrac{1}{3}\pi\right) = 2 \times \frac{\sqrt{3}}{2} = \sqrt{3}$$

Nugget – Graphical calculator notation

If you have access to a graphical calculator, you will discover that the notation used for substituting a numerical value into a function is very similar to that described above.

For example, in the first screenshot below, the function $y = x^2 - 4x + 3$ has been entered into the 'Y=' screen as the 'Y$_1$' function (i.e. the first function on the list).

Plot 1	Plot 2	Plot 3
\Y$_1$ = $x^2 - 4x + 3$		
\Y$_2$ =		
\Y$_3$ =		
\Y$_4$ =		
\Y$_5$ =		
\Y$_6$ =		
\Y$_7$ =		

Figure 1.1a

In the second screenshot, the user has returned to the home screen, entered Y$_1$(2) and the value of this function has been evaluated for $x = 2$.

Y$_1$(2)	
	−1

Figure 1.1b Screenshots taken from a TI-84 Plus graphical calculator.

1.5 Notation for increases in functions

If x is a variable, the symbol δx (or sometimes Δx) is used to denote an increase in the value of x. A similar notation is used for other variables. The symbol 'δ' is the Greek small 'd', called delta. Contrary to the usual notation in algebra, δx does not mean $\delta \times x$. In this case, the letters must not be separated. Thus, δx means an increase in x.

If y is a function of x, and x increases by δx, then y will increase as a result. This increase in y is denoted by δy.

Therefore if $$y = f(x)$$
then $$y + \delta y = f(x + \delta x)$$
so $$\delta y = f(x + \delta x) - f(x)$$

Notice that an increase could be negative as well as positive. It is possible when x is increased by 1 that y could be decreased by 1. In this case the increase $\delta x = 1$ and the increase $\delta y = -1$. Similarly, the increase in δx could be negative. The word 'increase' will be used in this deliberately ambiguous way throughout the book. If the increase is actually a decrease, its sign will be negative.

Example 1.5.1

Let $y = x^3 - 7x^2 + 8x$.

If x is increased by δx, y will be increased by δy. Then

$$y + \delta y = (x + \delta x)^3 - 7(x + \delta x)^2 + 8(x + \delta x)$$

Example 1.5.2

If $s = ut + \dfrac{1}{2}at^2$ and t is increased by δt, then s will be increased by δs, and

$$s + \delta s = u(t + \delta t) + \frac{1}{2}a(t + \delta t)^2$$

Sometimes single letters are used to denote increases, instead of the delta notation.

Example 1.5.3

Let $y = f(x)$. Let x be increased by h and let the corresponding increase in y be k. Then

$$y + k = f(x + h)$$

and

$$k = f(x + h) - f(x)$$

1.6 Graphs of functions

Let $f(x)$ be a function of x. Then, by the definition of a function, in Section 1.2, for every value taken by x there is a corresponding value of $f(x)$. So, by giving a set of values to x you obtain a corresponding set of values for y. You can plot the pairs of values of x and $f(x)$ to give the graph of $f(x)$.

Nugget – What is *f(x)* on the graph?

When it comes to understanding the big ideas of mathematics, the simplest questions are often the most useful. One query that often crops up with students is how $f(x)$ can be interpreted on a graph. Suppose that you have drawn the graph of, say, $f(x) = 2x + 5$. It is important to remember here that $f(x)$ corresponds to the value of y. So, if you were to choose a particular point on the graph of this function and consider the coordinates of that point, the x-coordinate will represent its horizontal distance from the y-axis, and its value for $f(x)$ (i.e. the y-coordinate) corresponds to the height of the point above the x-axis.

Example 1.6.1

Consider $f(x) = x^2$ or $y = x^2$. Giving x the values 0, 1, 2, 3, –1, –2, –3, ... you obtain the corresponding values of $f(x)$ or y.

Thus

$$f(0) = 0,$$

$$f(1) = 1, f(-1) = 1,$$

$$f(2) = 4, f(-2) = 4,$$

$$f(3) = 9, f(-3) = 9.$$

You can deduce that, for any value of a, $f(-a)$ has the same value as $f(a)$, so the curve must be symmetrical about the y-axis. The curve, a parabola, is shown in Figure 1.2.

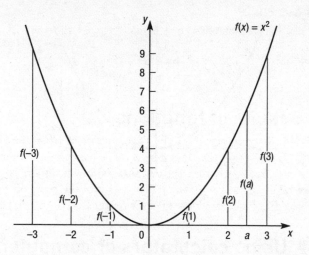

Figure 1.2 The graph of $f(x) = x^2$

At the points on the x-axis where $x = 1, 2, 3, \ldots$ the corresponding lines are drawn parallel to the y-axis to meet the graph. These lines are called **ordinates**. The lengths of these ordinates are $f(1), f(2), f(3), \ldots$, and the length of the ordinate drawn where $x = a$ is $f(a)$.

In Figure 1.3, which is part of the graph of $f(x) = x^2$ or $y = x^2$, the points L and N are taken on the x-axis so that

$$OL = a, ON = b$$

The lengths of the corresponding ordinates KL, MN are given by

$$KL = f(a), MN = f(b)$$

In general, let L be any point on the x-axis with $OL = x$. Increase x by LN where $LN = \delta x$. The corresponding increase in $f(x)$ or y is MP.

Therefore
$$MP = \delta y$$

As
$$KL = f(x) \quad \text{and} \quad MN = f(x + \delta x)$$
$$MP = f(x + \delta x) - f(x)$$

or
$$\delta y = f(x + \delta x) - f(x)$$

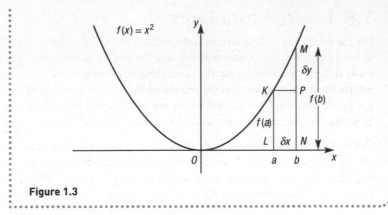

Figure 1.3

1.7 Using calculators or computers for plotting functions

You can use graphics calculators or computers to gain considerable insight into functions, and into the ideas of the calculus.

If you know a formula for a function $f(x)$, you can use a graphics calculator to calculate values of $f(x)$ and to draw the graph of $y = f(x)$ over any interval of values of x and y that you choose. This set of values for x and y is often called a **window**. Different calculators behave in different ways, so you must look in your manual to see how to draw the graph of, for example, $y = x^2$ on your calculator.

In particular, you will find it is useful to use the zoom function, so that you can magnify and look more closely at a part of the curve.

Similarly, you can use a computer for drawing graphs, with a graphing spreadsheet such as Excel, or any other graph-plotting software. Using Excel, you will need first to create a table of about 30 values for the function, and then, from the table, use the scatterplot graphing facility to draw a conventional graph of the function. Once again you can look at any part of the graph that you wish, and you should be able to look closely at a particular part of the graph by adjusting the scales.

1.8 Inverse functions

For the function $y = 2x$, you can deduce that $x = \frac{1}{2}y$. The graphs of $y = 2x$ and $x = \frac{1}{2}y$ are identical, because the two equations are really different ways of saying the same thing. However, if you reverse the roles of x and y in the second of them to get $y = 2x$ and $y = \frac{1}{2}x$, the functions $y = 2x$ and $y = \frac{1}{2}x$ are called **inverse** functions. You can think of each of them as undoing the effect of the other.

Similarly, for the function $y = x^2$ where $x \geq 0$, you can deduce that $x = \sqrt{y}$, where $y \geq 0$ and $\sqrt{y}$ means the positive value which squares to give y. The two functions, $y = x^2$ where $x \geq 0$, and $y = \sqrt{x}$ where $y \geq 0$, are also inverse functions.

1.9 Implicit functions

A function of the form $y = x^2 + 3x$ in which you are told directly the value of y in terms of x is called an **explicit** function. Thus $y = \sin x$ is an explicit function, but $x + y = 3$ is not, because the function is not in the explicit form starting with '$y =$'.

If an equation such as

$$x^2 - 2xy - 3y = 4$$

can be satisfied by values of x and y, but x and y are together on the same side of the equation, then y is said to be an **implicit** function of x. In this particular case you can solve for y in terms of x, giving

$$y = \frac{x^2 - 4}{2x + 3}$$

which is an explicit function for y.

Further examples of implicit functions are $x^3 - 3x^2y + 5y^3 - 7 = 0$ and $x \sin y + y^2 = 4xy$.

1.10 Functions of more than one variable

The functions you have met so far in this book have all been functions of one variable, but you can have functions of two or more variables.

For example, the area of a triangle is a function of both its base and height; the volume of a fixed mass of gas is a function of its pressure and temperature; the volume of a rectangular-shaped room is a function of the three variables: length, breadth and height.

This book, however, deals only with functions of one variable.

Test yourself

1 Let $f(x) = 2x^2 - 4x + 1$.
    Find the values of $f(1)$, $f(0)$, $f(2)$, $f(-2)$, $f(a)$ and $f(x + \delta x)$.

2 Let $f(x) = (x - 1)(x + 5)$.
    Find the values of $f(2)$, $f(1)$, $f(0)$, $f(a + 1)$, $f(\frac{1}{a})$ and $f(-5)$.

3 Let $f(\theta) = \cos \theta$.
    Find the values of $f(\frac{1}{2}\pi)$, $f(0)$, $f(\frac{1}{3}\pi)$, $f(\frac{1}{2}\pi)$ and $f(\pi)$.

4 Let $f(x) = x^2$.
    Find the values of $f(3)$, $f(3.1)$, $f(3.01)$ and $f(3.001)$.

    Also find the value of $\dfrac{f(3.001) - f(3)}{0.001}$

5 Let $\phi(x) = 2^x$.
    Find the values of $\phi(0)$, $\phi(1)$, $\phi(3)$ and $\phi(0.5)$.

6 Let $f(x) = x^3 - 5x^2 - 3x + 7$.
    Find the values of $f(0)$, $f(1)$, $f(2)$ and $f(-x)$.

7 Let $f(t) = 3t^2 + 5t - 1$.
    Find an expression for $f(t + \delta t)$.

8 Let $f(x) = x^2 + 2x + 1$.
    Find and simplify an expression for $f(x + \delta x) - f(x)$.

9 Let $f(x) = x^3$.
    Find expressions for the following:

    a $f(x + \delta x)$ **b** $f(x + \delta x) - f(x)$ **c** $\dfrac{f(x + \delta x) - f(x)}{\delta x}$

10 Let $f(x) = 2x^2$. Find expressions for the following:

    a $f(x + h)$ **b** $f(x + h) - f(x)$ **c** $\dfrac{f(x + \delta x) - f(x)}{\delta x}$

Key ideas

▶ Functions are expressed using a combination of **variables** (written with letters) and **constants** (typically written with **numbers**, but note that π, e and the gravitational constant g are constants).

▶ A **function** states the connection between two or more variables.

▶ An important distinction is that between **dependent** and **independent** variables (if a variable's value depends on that assigned to another, it is a dependent variable).

▶ Two common notations for referring to a function are of the form '$y = ...$' and '$f(x) = ...$'

▶ The symbol δ (pronounced delta) means 'a small change in' – for example δx is a small change in x.

▶ Functions that 'undo' each other are called **inverse** functions (for example, doubling is the inverse function for halving, and vice versa).

▶ An **explicit** function might be of the form $y =$ some expression in x, whereas an **implicit** function would have the x and y components mixed up together on the same side of the equation.

2

Variations in functions; limits

In this chapter you will learn:

▶ *the meaning of a limit*
▶ *how to deal with limits of the form $\frac{0}{0}$*
▶ *some simple theorems on limits, and how to use them.*

2.1 Variations in functions

When the independent variable in a function changes in value, the dependent variable changes its value as a result. Here are some examples which show how the dependent variable changes when the independent variable changes.

Example 2.1.1

Consider the function $f(x) = x^2$ or $y = x^2$ and refer to Figure 1.2. It shows how the function changes when x changes.

Remembering that the values of x are shown as continuously increasing from left to right, you can see the following features from the graph.

▶ As x increases continuously through negative values to zero, the values of y are positive and decrease to zero.

▶ As x increases through positive values, y also increases and is positive.

▶ At the origin, y ceases to decrease and starts to increase. This is called a **turning point** on the curve.

▶ If x is increased without limit, y will increase without limit. For values of x which are negative, but which are numerically very large, y is also very large and positive.

Nugget – Reciprocals

In the example below, you will use the function $y = \frac{1}{x}$, which is called the **reciprocal function**. As can be seen from its equation, the y-value is found by dividing 1 by the corresponding x-value. Most calculators, even the simpler models, possess a reciprocal key, usually marked $\frac{1}{x}$ or x^{-1}. Note that a special feature of reciprocals is that by applying this function twice in succession to a number, it returns to its starting value.

Example 2.1.2

Consider the function $y = \dfrac{1}{x}$

For this function, you can see the following features.

▶ if x is very large, say 10^{10}, y is a very small number

▶ if $x = 10^{100}$, $y = \frac{1}{10^{100}}$, which is very small indeed.

These numbers, both very large and very small, are numbers which you can write down. They are **finite** numbers.

If, however, you increase x so that it eventually becomes bigger than any finite number, however large, which you might choose, then x is said to be **increasing without limit**. It is said to approach infinity; you write this as $x \to \infty$.

You should note that ∞ is not a number: you cannot carry out the usual arithmetic operations on it.

Similarly, as x becomes very large, the function $\dfrac{1}{x}$ becomes very small. In fact, as $x \to \infty$, $\dfrac{1}{x}$ becomes smaller in magnitude than any number, however small, you might choose.

In this case, $\dfrac{1}{x}$ is said to approach 0; this is written as $\dfrac{1}{x} \to 0$

Thus, when
$$x \to \infty, \frac{1}{x} \to 0$$

Also, when x is positive, as
$$x \to 0, \frac{1}{x} \to \infty$$

Notice that the same conclusions hold if the numerator is any positive number a.

Therefore, when x is positive as $x \to 0$, $\dfrac{a}{x} \to \infty$

These results are illustrated in the graph of $y = \dfrac{1}{x}$ shown in Figure 2.1.

The curve is called a hyperbola, and consists of two branches of the same shape, corresponding to positive and negative values of x.

From the positive branch, you can see the graphical interpretation of the conclusions reached above.

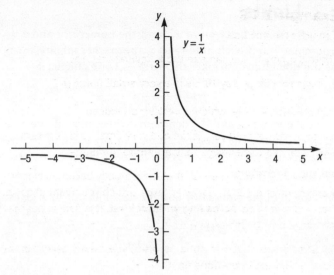

Figure 2.1 The graph of $y = \frac{1}{x}$

▶ As x increases, y decreases and the curve approaches the x-axis. As x approaches infinity, the distance between the curve and the x-axis becomes very small and the curve approaches the x-axis as the value of x approaches infinity. Geometrically, you could say that the x-axis is tangential to the curve at infinity.

▶ For values between 0 and 1, note that the curve approaches the y-axis as x approaches 0, that is, the y-axis is tangential to the curve at infinity.

A straight line which is tangential to a curve at infinity, is called an **asymptote** to the curve.

Thus the axes are both **asymptotes to the curve** $y = \dfrac{1}{x}$

The arguments above apply also to the branch of the curve corresponding to negative values of x. Both axes are asymptotes to the curve in negative directions.

Notice also that throughout the whole range of numerical values of x, positive and negative, y is always decreasing. The sudden change as x passes through zero is a matter for consideration later.

2.2 Limits

If in a fractional function of x, both the numerator and denominator approach infinity as x approaches infinity, then the fraction ultimately takes the form $\frac{\infty}{\infty}$. For example, if

$$f(x) = \frac{2x}{x+1}$$

both the numerator and the denominator become infinite as x becomes infinite. The question arises, can any meaning be given to the fraction when it assumes the form $\frac{\infty}{\infty}$? Here is a way to find such a meaning.

Dividing both the numerator and the denominator by x gives

$$f(x) = \frac{2x}{x+1} = \frac{2}{1+\dfrac{1}{x}}$$

If now, $x \to \infty$, then $\dfrac{1}{x} \to 0$, so the fraction approaches $\dfrac{2}{1+0} = 2$

But clearly, the fraction cannot be bigger than 2, so $\dfrac{2x}{x+1}$ approaches the limiting value 2 as x approaches infinity; it is called the **limit of the function**. The limit of this function is denoted by

$$\lim_{x \to \infty} \frac{2x}{x+1} = 2$$

The value towards which x approaches when the limit is found, is indicated by the $x \to \infty$ placed beneath the word 'lim'.

The idea of a limit is not only fundamental to calculus, but also to nearly all advanced mathematics.

2.3 Limit of a function of the form $\frac{0}{0}$

Consider the function $f(x) = \dfrac{x^2-4}{x-2}$

You can find the value of this function for every value of x except 2. If x is given the value 2, both numerator and denominator become 0, and the function is not defined there. But when x is close to 2, the function takes the form $\frac{0}{0}$. This form is called **indeterminate**.

Nugget – Dividing by zero

Try dividing any number by zero on your calculator or computer and you should get some sort of error message. A well-designed machine will actually display, 'ERROR: DIVIDE BY 0'. Alternatively you might just be offered any of the following: 'MATH ERROR'; 'UNDEFINED'; or simply '-E-'.

The form $\frac{0}{0}$ is of great importance and needs further investigation. Start by giving x values which are slightly greater or slightly less than the value which produces the indeterminate form, that is 2.

Using a calculator, you find the following results.

Value of x	Value of $\dfrac{x^2-4}{x-2}$	Value of x	Value of $\dfrac{x^2-4}{x-2}$
2.1	4.1	1.9	3.9
2.01	4.01	1.99	3.99
2.001	4.001	1.999	3.999
2.0001	4.0001	1.9999	3.9999

A comparison of these results leads to the conclusion that, as the value of x approaches 2, the value of the fraction approaches 4. You can make the value of the fraction as close to 4 as you choose by taking x to be close enough to 2. This can be expressed in the form as

$$x \to 2, \frac{x^2-4}{x-2} \to 4$$

The function $\dfrac{x^2-4}{x-2}$ has the limiting value 4 as x approaches 2, so, with the notation for limits

$$\lim_{x \to 2} \frac{x^2-4}{x-2} = 4$$

Example 2.3.1

Now consider a more general problem, namely the value of $\dfrac{x^2-a^2}{x-a}$ as $x \to a$. The method used is similar to the one above, but in a more general form. Notice that you cannot put x equal to a.

Let $x = a + h$, that is, h is the amount by which x differs from a.

Substituting in the fraction

$$\frac{x^2 - a^2}{x - a} = \frac{(a + h)^2 - a^2}{(a + h) - a}$$

$$= \frac{2ah + h^2}{h}$$

Dividing the numerator and denominator by h, which cannot be zero,

$$\frac{x^2 - a^2}{x - a} = 2a + h$$

When x approaches a, h approaches 0, so $2a + h$ approaches $2a$. That is, as x approaches a, $\dfrac{x^2 - a^2}{x - a}$ approaches $2a$, or alternatively, $2a$ is the limiting value of the function.

Using limit notation:

$$\lim_{x \to a} \frac{x^2 - a^2}{x - a} = 2a$$

You can regard the ratio $\frac{0}{0}$, as used in the example above, as the ratio of two very small magnitudes. The value of this ratio approaches a limit as the numerator and the denominator approach zero.

2.4 A trigonometric limit, $\lim\limits_{\theta \to 0} \dfrac{\sin\theta}{\theta} = 1$

Throughout this book, angles will always be measured in radians, unless specified to the contrary.

It is clear that, as θ becomes very small, so also does $\sin \theta$, so that the ratio $\frac{\sin\theta}{\theta}$ is of the form $\frac{0}{0}$ as θ approaches 0.

You can find the limit of this ratio as follows. In Figure 2.2, let O be the centre of a circle of radius 1 unit. Let OA be a radius in the x-direction and let $\angle AOB$ be θ radians. The perpendicular from B to OA meets OA at D, and the tangent to the circle at A meets OB produced at T.

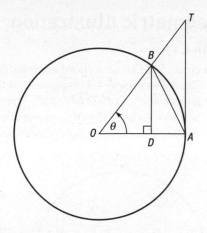

Figure 2.2

From Figure 2.2, area $\triangle OAB$ < area sector OAB < area $\triangle OAT$

The area of $\triangle OAB = \frac{1}{2}OA \times OB \times \sin\theta = \frac{1}{2} \times 1 \times 1 \times \sin\theta$

The area of sector $OAB = \frac{1}{2} \times OA^2 \times \theta = \frac{1}{2} \times 1 \times \theta$

The area of $\triangle OAT = \frac{1}{2}OA \times AT = \frac{1}{2} \times 1 \times \tan\theta$

Substituting these expressions in the inequality above:

$$\frac{1}{2}\sin\theta < \frac{1}{2}\theta < \frac{1}{2}\tan\theta \ \ or \ \ \sin\theta < \theta < \frac{\sin\theta}{\cos\theta}$$

Dividing throughout by $\sin\theta$ gives: $1 < \dfrac{\theta}{\sin\theta} < \dfrac{1}{\cos\theta}$

But as $\theta \to 0$, $\cos\theta \to 1$, so $\dfrac{1}{\cos\theta} \to 1$

But $\dfrac{\theta}{\sin\theta}$ is always sandwiched between 1 and $\dfrac{1}{\cos\theta}$

Therefore as $\theta \to 0$, $\dfrac{\theta}{\sin\theta} \to 1$.

In limit notation, $\displaystyle\lim_{\theta \to 0}\dfrac{\theta}{\sin\theta} \to 1$.

2.5 A geometric illustration of a limit

Let *OAB* be a circle. Let *OB* be a chord intersecting the circle at *O* and *B*, as shown in Figure 2.3. Suppose that the chord *OB* rotates in a clockwise direction about *O*. The point of intersection *B* will move about the circumference towards *O*. Consequently, the arc *OB* and the chord *OB* decrease.

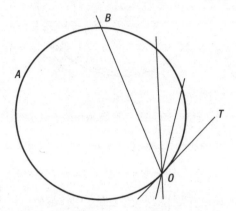

Figure 2.3

Let the rotation continue until *B* is very close indeed to *O*, and the chord and the arc both become very small indeed.

When the point *B* eventually coincides with *O*, the chord *OB* does not cut the circle at a second point. Therefore, in the limiting position the chord becomes a tangent to the circle at *O*.

2.6 Theorems on limits

Here are statements, without proofs, of four theorems on limits. You can omit them, if you wish, when you first read the book.

1 Limit of equal quantities

If two variables are always equal, their limits are equal.

2 Limit of a sum

The limit of the sum of any number of functions is equal to the sum of the limits of the separate functions.

That is, if u and v are functions of x, then

$$\lim (u + v) = \lim u + \lim v$$

3 Limit of a product

The limit of the product of any number of functions is equal to the product of the limits of the separate functions.

That is, if u and v are functions of x, then

$$\lim (u \times v) = \lim u \times \lim v$$

4 Limit of a quotient

The limit of the quotient of two functions is equal to the quotient of the limits of the separate functions, provided that the limit of the divisor is non-zero.

That is, if u and v are functions of x, then

$$\lim \left(\frac{u}{v} \right) = \frac{\lim u}{\lim v}, \text{ provided } \lim v \neq 0$$

Example 2.6.1

Find the limit of $\dfrac{x^2 + 3x}{2x^2 - 5}$ as x becomes infinite.

$$\lim_{x \to \infty} \frac{x^2 + 3x}{2x^2 - 5} = \lim_{x \to \infty} \frac{1 + \dfrac{3}{x}}{2 - \dfrac{5}{x^2}}$$

$$= \frac{1 + \lim\limits_{x \to \infty} \dfrac{3}{x}}{2 - \lim\limits_{x \to \infty} \dfrac{5}{x^2}} \quad \text{(by Theorem 4)}$$

$$= \frac{1 + 0}{2 + 0} = \frac{1}{2}$$

Example 2.6.2

Find the value of $\lim\limits_{x \to a} \dfrac{x^n - a^n}{x - a}$, where n is a positive integer.

When $x = a$, the function is of the form $\frac{0}{0}$, and is indeterminate.

Let $x = a + h$. Then

$$\frac{x^n - a^n}{x - a} = \frac{(a + h)^n - a^n}{(a + h) - a}$$

Expanding $(a + h)^n$ by the binomial theorem:

$$\frac{x^n - a^n}{x - a} = \frac{\left\{ a^n + na^{n-1}h + \dfrac{n(n-1)}{2!} a^{n-2}h^2 + \cdots \right\} - a^n}{h}$$

$$= \left\{ na^{n-1} 1 + \frac{n(n-1)}{2!} a^{n-2}h + \cdots \right\}$$

But since $x = a + h$, when $x \to a$, $h \to 0$

Therefore

$$\lim_{x \to a} \frac{x^n - a^n}{x - a} = \lim_{h \to 0} \left\{ na^{n-1} + \frac{n(n-1)}{2!} a^{n-2}h + \cdots \right\}$$

$$= na^{n-1}$$

since all the other terms have a power of h as a factor, and therefore vanish when $h \to 0$.

Example 2.6.3

Find the limit of $\dfrac{x - 3}{\sqrt{x - 2} - \sqrt{4 - x}}$ as $x \to 3$

Both the numerator and the denominator are zero when $x = 3$, so the function takes the form $\frac{0}{0}$, and is indeterminate.

Multiplying the numerator and denominator by $\sqrt{x-2}+\sqrt{4-x}$ gives:

$$\frac{x-3}{\sqrt{x-2}-\sqrt{4-x}}=\frac{(x-3)}{\sqrt{x-2}-\sqrt{4-x}}\times\frac{\sqrt{x-2}+\sqrt{4-x}}{\sqrt{x-2}+\sqrt{4-x}}$$

$$=\frac{(x-3)(\sqrt{x-2}+\sqrt{4-x})}{2x-6}$$

$$=\frac{(\sqrt{x-2}+\sqrt{4-x})}{2}$$

so

$$\lim_{x\to3}\frac{x-3}{\sqrt{x-2}-\sqrt{4-x}}=\frac{\sqrt{1}+\sqrt{1}}{2}=1$$

Test yourself

1 a What number does the function $f(x)=\dfrac{1}{x-1}$ approach as x becomes infinitely large?

  b For what values of x is the function negative?

  c What are the values of the function when the values of x are 2, 1.8, 1.5, 1.2, 1.1, 0.5, 0, −1, −2?

  d What limit is approached by the function as x approaches 1?

  e Using the values of the function found in part (**c**) sketch the graph of the function.

2 a Find the values of the function $f(x)=\dfrac{3x+1}{x}$

  when x has the values 10, 100, 1000, 1 000 000.

  b What limit does the function approach as x becomes very great?

  c Find the limit of the function as x becomes infinitely large by using the method of Example 2.3.1.

3 a Find $\lim\limits_{x\to\infty}\dfrac{5x+2}{x-1}$.

  b Find the limit of the function as x approaches 1.

4 a Find the values of the function $\dfrac{x^2-1}{x-1}$

when x has the values 10, 4, 2, 1.5, 1.1, 1.01.

b Find the limit of $\dfrac{x^2-1}{x-1}$ as x approaches 1.

5 Find the limit of the function $\dfrac{x^2-1}{x-1}$ as x approaches infinity.

6 Find $\displaystyle\lim_{x\to 2}\dfrac{x^2-4}{x^2-2x}$

7 Find the limit of $\dfrac{(x+h)^3-x^3}{h}$ as $h\to 0$

8 Find the limit of the function $\dfrac{x}{2x+1}$ as x approaches ∞.

9 Find $\displaystyle\lim_{x\to\infty}\dfrac{4x^2+x+1}{3x^2+2x+1}$

10 Show from the proof given in Section 2.4 that $\displaystyle\lim_{\theta\to 0}\dfrac{\tan\theta}{\theta}=1$

Key ideas

▶ When the independent variable in a function changes in value, the dependent variable changes its value as a result.

▶ An *asymptote* is a straight line that is tangential to a curve (for example, the two axes are asymptotes to the curve $y=\frac{1}{x}$).

▶ The *limiting value* of a function is its value as the value of x approaches infinity.

3

Gradient

In this chapter you will learn:

▶ *how to find the gradient of a straight line*
▶ *how to use the gradient of a straight line to define the gradient of a curve*
▶ *how to interpret positive and negative gradients in terms of increasing and decreasing functions.*

3.1 Gradient of the line joining two points

The gradient of the straight line joining two points A and B is a measure of the slope of the line. In Figure 3.1 the point A is $(1, 2)$ and the point B is $(4, 4)$.

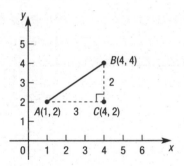

Figure 3.1

Draw the line from A parallel to the x-axis to meet the line drawn from B parallel to the y-axis at C. Then the coordinates of C are $(4, 2)$. The gradient of the line AB is defined to be the distance CB divided by the distance AC, that is:

$$\text{gradient } AB = \frac{CB}{AC} = \frac{4-2}{4-1} = \frac{2}{3}$$

Definition: The gradient of the straight line joining two points $P(x_1, y_1)$ and $Q(x_2, y_2)$ is defined as:

$$\text{gradient } PQ = \frac{y_2 - y_1}{x_2 - x_1}$$

You can interpret this equation by saying that the gradient of the straight line joining two points P and Q is the y-increase from P to Q, divided by the x-increase from P to Q (see Figure 3.2a).

Using the notation of Section 1.5, if the y-increase from P to Q is δy, and the x-increase from P to Q is δx (see Figure 3.2b), then

$$\text{gradient } PQ = \frac{\delta y}{\delta x}$$

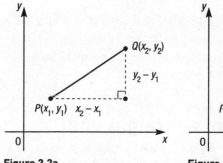

Figure 3.2a   Figure 3.2b

Example 3.1.1

Find the gradient of the straight line joining $(2, 3)$ to $(5, 2)$.

Using the formula:

$$\text{gradient } PQ = \frac{y_2 - y_1}{x_2 - x_1}$$

you find that:

$$\text{gradient } PQ = \frac{2-3}{5-2} = \frac{-1}{3} = -\frac{1}{3}$$

Thus the gradient of a straight line is negative if it slopes downwards as x increases, as below.

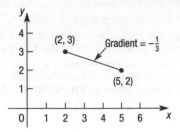

Figure 3.3

Notice also that it doesn't matter which point you take as *P* and which as *Q* in the formula. If you take them the other way round you find that:

$$\text{gradient } PQ = \frac{3-2}{2-5} = \frac{1}{-3} = -\frac{1}{3}$$

It is important to realize that the 'y-increase' corresponding to the x-increase can actually be a decrease.

3.2 Equation of a straight line

A straight line is a curve with a constant gradient. Suppose that a straight line with gradient *m* passes through the point *A* with coordinates (*a*, *b*) (see Figure 3.4).

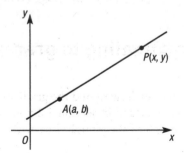

Figure 3.4

Let *P*(*x*, *y*) be the coordinates of any point on the straight line. Then, using the result of Section 3.1:

$$\text{gradient } AP = \frac{y-b}{x-a} = m$$

Rewriting this equation gives:

$$y - b = m(x - a) \qquad \textbf{Equation 1}$$

You can also rewrite this equation in the form $y = mx + c$, where c is a constant with value $c = b - ma$.

Thus the equation $y = mx + c$ represents a straight line of gradient m.

Example 3.2.1

Find the equation of the straight line through $(1, 2)$ with gradient -3.

Using **equation 1**, the straight line has equation $y - 2 = -3(x - 1)$, that is, $y = -3x + 5$.

Example 3.2.2

Find the coordinates of the point at which the straight line $y = mx + c$ meets the y-axis.

The line $y = mx + c$ meets the y-axis at the point for which $x = 0$. Thus, substituting $x = 0$, you find that $y = c$. Therefore the coordinates of the point are $(0, c)$.

This point is sometimes called the **intercept on the y-axis**.

3.3 Approximating to gradients of curves

In Section 3.1, you saw how to find the gradient of the straight line joining two points, and in Section 3.2 you saw that the gradient of the graph of the straight line with equation $y = mx + c$ is m.

If the graph is a curve, it is not obvious what is meant by its gradient, as it is always changing. Nor is it obvious what the gradient at a point on the curve means.

However, you can get a good impression of the gradient of a curve by drawing the graph on a graphics calculator and zooming in on the graph, looking only at a small part. Figure 3.5 shows the graph of $y = x^2$ and an enlarged part of the curve close to the point $x = 1$, $y = 1$.

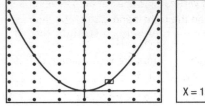 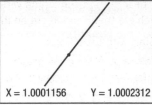

Figure 3.5

You can see that the curve now *looks like* a straight line. So you can find an approximation to the gradient of the curve by finding the gradient of that straight line, using the coordinates of two points on it. Since the point (1, 1) lies on the graph, you can use this as one of the points. You can find the coordinates of another point on the curve in the graph window by using the trace function. In the case of Figure 3.5, the point taken was (1.000 1156, 1.000 2312). Using the result in Section 3.1 for finding the gradient of the line joining two points, they give

$$\frac{1.000\,2312 - 1}{1.000\,1156 - 1} = \frac{0.000\,2312}{0.000\,1156} = 2.000\,116\ldots$$

as an approximation to the gradient of the curve.

In general if you zoom in further, you will get a better approximation.

3.4 Towards a definition of gradient

You can see from Section 3.3 that the gradient at (1, 1) on the graph of $y = x^2$ is approximately 2. Unfortunately, however, this really isn't good enough. But it does begin to point the way to a definition of gradient which will enable you to calculate an exact answer.

The following examples give an indication of the method. A definition of gradient follows in the next section.

Example 3.4.1

Find the gradient at (1, 1) on the graph of $y = x^2$.

On the left-hand side of the vertical line is a set of instructions which are then carried out on the right-hand side of the line.

Start with the point A.	A is $(1, 1)$
Take a point P on the graph such that the x-increase is δx.	Thus the x-coordinate of P is $1 + \delta x$
Find the y-coordinate of P by substituting in the equation $y = x^2$.	The y-coordinate of P is $(1 + \delta x)^2$ or, after removing the bracket, $1 + 2\delta x + (\delta x)^2$
Find the y-increase, δy, from A to P.	$\delta y = 1 + 2\delta x + (\delta x)^2 - 1$ $= 2\delta x + (\delta x)^2$
Calculate the gradient of AP.	$\text{Gradient } AP = \dfrac{\delta y}{\delta x}$ $= \dfrac{2\delta x + (\delta x)^2}{\delta x}$ $= 2 + \delta x$

So far, nothing in this calculation has corresponded to the zooming process in which the value of δx gets progressively smaller. You can do this by using the idea of a limit from Chapter 2. If you now let $\delta x \to 0$, the chord becomes indistinguishable from the tangent at the point $x = 1$, so the gradient of the chord is very close to the gradient of the tangent to the curve.

But $\lim\limits_{\delta x \to 0} \dfrac{\delta y}{\delta x} = \lim\limits_{\delta x \to 0} (2 + \delta x) = 2$, so it is reasonable to conclude that the gradient of the tangent at $x = 1$ is 2.

Example 3.4.2

Find the gradient of $y = x^2$ at $x = 2$.

Let the point with an x-coordinate of 2 be A, so that the coordinates of A are $(2, 4)$. Let P have x-coordinate $2 + \delta x$. The y-coordinate of P is $(2 + \delta x)^2$ or, after multiplying out the bracket, $4 + 4\delta x + (\delta x)^2$. The y-increase, δy, is given by the equation

$$\delta y = 4 + 4\delta x + (\delta x)^2 - 4 = 4\delta x + (\delta x)^2$$

Then

$$\text{gradient } AP = \frac{\delta y}{\delta x} = \frac{4\delta x + (\delta x)^2}{\delta x} = 4 + \delta x$$

Therefore the gradient at $x = 2$ is

$$\lim_{\delta x \to 0} \frac{\delta y}{\delta x} = \lim_{\delta x \to 0} (4 + \delta x) = 4$$

Example 3.4.3

Find the gradient of $y = x^2$ at $x = a$.

Following the method of Examples 3.4.1 and 3.4.2, the point A is (a, a^2), and P has an x-coordinate of $a + \delta x$. The y-coordinate of P is $(a + \delta x)^2$ or, after multiplying out the bracket, $a^2 + 2a\delta x + (\delta x)^2$. The y-increase, δy, is given by the equation

$$\delta y = a^2 + 2a\delta x + (\delta x)^2 - a^2 = 2a\delta x + (\delta x)^2$$

Then

$$\text{gradient } AP = \frac{\delta y}{\delta x} = \frac{2a\delta x + (\delta x)^2}{\delta x} = 2a + \delta x$$

Therefore the gradient of the tangent at $x = a$ is

$$\lim_{\delta x \to 0} \frac{\delta y}{\delta x} = \lim_{\delta x \to 0} (2a + \delta x) = 2a$$

3.5 Definition of the gradient of a curve

Following the work in Examples 3.4.1 to 3.4.3, it is now possible to give a definition of the gradient at a point on a curve.

In fact two definitions are given. They differ only in notation.

Definition 1: The gradient at a point on a curve $x = a$ is:

$$\lim_{\delta x \to 0} \frac{\delta y}{\delta x}$$

where δx and δy are the increases from the point $x = a$ to another point on the graph.

You can modify this definition to use function notation.

Following the same procedures as in Examples 3.4.1 to 3.4.3, the point A has coordinates $(a, f(a))$ and P has x-coordinate $a + \delta x$. The y-coordinate of P is $f(a + \delta x)$. The y-increase, δy, is given by the equation $\delta y = f(a + \delta x) - f(a)$.

Then:
$$\text{gradient } AP = \frac{\delta y}{\delta x} = \frac{f(a + \delta x) - f(a)}{\delta x}$$

Definition 2: The gradient of the tangent at $x = a$ is

$$\lim_{\delta x \to 0} \frac{\delta y}{\delta x} = \lim_{\delta x \to 0} \frac{f(a + \delta x) - f(a)}{\delta x}$$

If the x-increase is h, then the gradient of the tangent at $x = a$ is

$$\lim_{h \to 0} \frac{f(a + h) - f(a)}{h}$$

3.6 Negative gradient

You can use the result of Example 3.4.3 to give the value of the gradient at any point on the graph of $y = x^2$. If you substitute $a = 1$ or $a = 2$, you obtain the results in the working before Example 3.4.1, in Example 3.4.1 itself and Example 3.4.2.

If you put $a = -1$, you find that the gradient is -2, which means that the graph of the function slopes downwards as you move from left to right through the point $x = -1$.

When the tangent has a negative gradient, the function is said to be **decreasing**. Similarly, when the tangent has a positive gradient, the function is said to be **increasing**.

Thus the function $f(x) = x^2$ decreases as x increases from large negative values to 0, and increases as x increases from 0 to large positive values.

These ideas will be taken further in Chapter 5.

Test yourself

1 Draw the graph of the straight line $3x - 2y = 6$, and find its gradient. Let P and Q be two points on the line such that the x-value of Q is 0.8 greater than the x-value of P. By how much is the y-value of Q greater than the y-value of P?

2 Find the gradients of the following straight lines.

    a $\frac{1}{2}x - \frac{1}{5}y = 4$ **b** $4x + 5y = 16$ **c** $\frac{x}{a} + \frac{y}{b} = 1$

3 The gradient of a straight line is 1.2. It passes through a point whose coordinates are (5, 10). Find the equation of the line.

4 For the curve with equation $y = x^2$, use the notation employed in Example 3.4.1 to find the value of $\frac{\delta y}{\delta x}$ as the value of x is increased from 3 to 3.1, 3.01, 3.001, 3.0001 respectively. Deduce the gradient of the tangent to the curve at the point where $x = 3$.

5 Draw the graph of $y = x^3$ for values of x between 0 and 2. (Use a graphics calculator if you have one.) Find an expression for δy in terms of δx for the point at which $x = a$. Hence find an expression for $\frac{\delta y}{\delta x}$ in terms of a and δx. Now let $a = 2$, and taking δx successively as 0.1, 0.01, 0.001 and 0.0001, find the limit which $\frac{\delta y}{\delta x}$ approaches as δx approaches 0.

6 For the function $y = \frac{1}{x}$, shown in Figure 2.1, find an expression for δy in terms of δx for the point at which $x = a$. Hence find an expression for $\frac{\delta y}{\delta x}$ in terms of a and δx. By taking the values of δx successively as 0.1, 0.01, 0.001 and 0.0001, find the limit which $\frac{\delta y}{\delta x}$ approaches as δx approaches 0 for the point $x = 1$.

   Hence find the gradient and angle of slope of the curve at the point for which $x = 1$. Check this result by using your graphics calculator, or by drawing.

7 Find the gradient of the tangent at $x = 1$ on the following curves.
    a $y = x^2 + 2$ **b** $y = x^2 - 3$

8 Find the gradient of the tangent at $x = 2$ on the following curves.
    a $y = 3x^2$ **b** $y = 2x^2 - 1$

Key ideas

▶ Calculus is fundamentally about the *rate of change* of a function with respect to the change in the variable on which it depends.

▶ *Uniform motion* means travelling over equal distances in equal intervals of time (i.e. where the velocity is constant).

▶ The *gradient* of a straight line is the tangent of the angle that the line makes with the positive horizontal axis. It is also the value of m in the equation $y = mx + c$.

▶ When y increases as x increases, the *gradient* is *positive*. When y decreases as x increases, the *gradient* is *negative*.

▶ The gradient of a curve at a point on the curve is equal to the gradient of the tangent to the curve at that point.

▶ If you consider a point on a curve, (x, y) and also a point very close to it $(x + \delta x, y + \delta y)$, the gradient of the chord joining the points is $\frac{\delta y}{\delta x}$.

4

Rate of change

In this chapter you will learn:

▶ *the meaning of average change of a function*

▶ *how to find actual velocity from average velocity*

▶ *a definition of 'rate of change'.*

4.1 The average change of a function over an interval

You saw in Section 2.1 that a function changes when the variable it depends on changes. But how can you measure the change or the rate of change of a function? This chapter is about measuring change.

The word 'rate' is part of everyday vocabulary. For example, the rate of change of prices is increasing. Or, prices are increasing at a dreadful rate. Or, the car is moving at a rate of 50 miles per hour. Rate is clearly something to do with change, but what precisely does it mean? Here is an example, in which the graph of a function is used to give a picture of the change in the function.

Example 4.1.1 Motion with constant velocity

Consider the following example. A motor car is moving with constant velocity 20 ms^{-1}. The distance covered and the time taken are both measured from a fixed point, which is taken as the origin, in the path of the motor car. Distance measured in this way is called **displacement**, and it may be positive or negative.

Time t seconds	0	1	2	3	4	5
Displacement s metres	0	20	40	60	80	100

The graph of these points is the straight line $s = 20t$ shown in Figure 4.1.

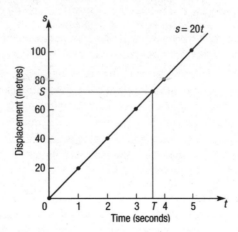

Figure 4.1

The average velocity over an interval of time is defined as:

$$\text{average velocity over the time interval} = \frac{\text{change in displacement}}{\text{change in time}}$$

Let T be any value of the time, and let the corresponding displacement be S. Then the average velocity over the time interval from the origin to time T is given by:

$$\frac{\text{change in displacement}}{\text{change in time}} = \frac{S}{T}$$

Since the values of S and T satisfy the equation of the line, $s = 20t$, you know that $S = 20T$, so that:

$$\text{average velocity for the interval} = \frac{\text{change in displacement}}{\text{change in time}} = \frac{S}{T} = 20$$

But $\frac{S}{T} = 20$ is also the gradient of the straight-line graph $s = 20t$, from the work of Section 3.1.

An important point arises from this example.

▶ For motion with constant velocity, the gradient of the straight line displacement-time graph represents the velocity.

As average velocity is a particular case of an average rate of change, here is a definition, based on the general idea of average rate of change.

Definition: The **average rate of change of a function** $f(x)$ is:

$$\begin{aligned}\text{average rate of change of } f(x) \atop \text{from } x_1 \text{ to } x_2 &= \frac{\text{change in } f(x)}{\text{change in } x} \\ &= \frac{f(x_2) - f(x_1)}{x_2 - x_1}\end{aligned}$$

Example 4.1.2

Find the average rate of change of the function $y = mx + c$.

Let P be any point on the line with an x-coordinate of a, and let the corresponding y-coordinate be b. Then, as P lies on the graph of $y = mx + c$, its coordinates must satisfy the equation $y = mx + c$.

Therefore $\qquad\qquad\qquad b = ma + c \qquad\qquad\qquad$ **Equation 1**

Let Q be another point on the graph, with x-coordinate $a + \delta x$, and let the corresponding y-coordinate be $b + \delta y$. As Q lies on the graph of $y = mx + c$, its coordinates must also satisfy the equation $y = mx + c$.

Therefore $\qquad\qquad b + \delta y = m(a + \delta x) + c$ $\qquad$ **Equation 2**

To find the average rate of change of y over an interval from $x = a$ to $x = a + \delta x$, you need to calculate the quotient

$$\frac{\text{increase in } f(x)}{\text{increase in } x} = \frac{f(x_2) - f(x_1)}{x_2 - x_1} = \frac{\delta y}{\delta x}$$

Subtracting Equation **1** from Equation **2** gives $\delta y = m(\delta x)$, so

$$\frac{\delta y}{\delta x} = m$$

Therefore the average rate of change of the linear function $y = mx + c$ over any interval of x whatsoever is m.

But the gradient of the straight line joining P and Q is $\dfrac{\delta y}{\delta x} = m$

▶ Therefore, for the linear function of which the graph is a straight line, the gradient of the line is the average rate of change of the function.

Nugget – Rise over run

You may remember the catchphrase mentioned in Chapter 3 for calculating a gradient, which was 'rise over run'. Using the notation above, the term *rise* corresponds to δy, run corresponds to δx and the gradient, or rise over run, is $\frac{\delta y}{\delta x}$.

4.2 The average rate of change of a non-linear function

Here is a generalization of the treatment of Example 4.1.2 which applies to a general function $y = f(x)$. Part of the graph of $y = f(x)$ is shown in Figure 4.2.

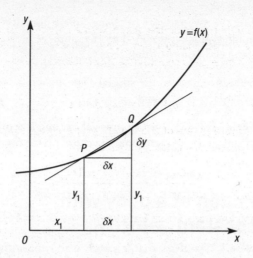

Figure 4.2

Let P be any point on the graph with an x-coordinate of x_1, and let the corresponding y-coordinate be y_1. Then, as P lies on the graph of $y = f(x)$, its coordinates must satisfy the equation $y = f(x)$.

Therefore $\qquad\qquad y_1 = f(x_1)$ $\qquad\qquad$ **Equation 3**

Let Q, with x-coordinate x_2, be another point on the graph and let the corresponding y-coordinate be y_2. Then, as Q lies on the graph of $y = f(x)$, its coordinates must also satisfy the equation $y = f(x)$.

Therefore $\qquad\qquad y_2 = f(x_2)$ $\qquad\qquad$ **Equation 4**

Suppose also that the increase, or step between x_1 and x_2 is δx, so that $x_2 - x_1 = \delta x$, and similarly $y_2 - y_1 = \delta y$.

The average rate of change of y over the interval from x_1 to x_2 is:

$$\frac{\text{increase in } f(x)}{\text{increase in } x} = \frac{f(x_2) - f(x_1)}{x_2 - x_1} = \frac{y_2 - y_1}{x_2 - x_1} = \frac{\delta y}{\delta x}$$

Looking now at Figure 4.2, the gradient of the straight line PQ joining the point (x_1, y_1) to (x_2, y_2) is also

$$\frac{y_2 - y_1}{x_2 - x_1} = \frac{\delta y}{\delta x}$$

▶ Therefore the average rate of change of the function $y = f(x)$ over an interval is represented by the gradient of the chord joining the end points of that interval.

Translating this remark to the special case of average velocity, gives the following conclusion.

▶ The average velocity over an interval of time is represented by the gradient of the corresponding chord on the displacement-time graph.

4.3 Motion of a body with non-constant velocity

In Example 4.1.1, you saw that if a body is moving with constant velocity, its displacement-time graph is a straight line. Now consider a body moving with increasing velocity. In this case, it is clear that in equal intervals of time, as the velocity increases, the displacements will also increase.

As an example, the equation $s = 5t^2$ is an approximation to the distance s in metres moved by a freely falling body t seconds after it is released from rest.

Suppose that you wanted to calculate the actual velocity at the instant when $t = 1$.

You can find the average velocity between the times $t = 1$ and $t = 2$ by using the formula:

$$\text{average velocity between times 1 and 2} = \frac{\text{increase in distance}}{\text{increase in time}}$$

$$= \frac{5 \times 2^2 - 5 \times 1^2}{2 - 1}$$

$$= \frac{20 - 5}{1} = 15 \text{ ms}^{-1}$$

Clearly this is not the actual velocity at $t = 1$, but an approximation to it. You could get a better approximation by taking the average velocity between $t = 1$ and $t = 1.5$:

Average velocity between times 1 and 1.5 $= \dfrac{\text{increase in distance}}{\text{increase in time}}$

$$= \frac{5 \times 1.5^2 - 5 \times 1^2}{1.5 - 1}$$

$$= \frac{11.25 - 5}{0.5} = 12.5 \text{ ms}^{-1}$$

This is still only an approximation to the actual velocity. For a better approximation, take the average velocity between $t = 1$ and $t = 1.1$.

Average velocity between times 1 and 1.1 $= \dfrac{\text{increase in distance}}{\text{increase in time}}$

$$= \frac{5 \times 1.1^2 - 5 \times 1^2}{1.1 - 1}$$

$$= \frac{6.05 - 5}{0.1} = 10.5 \text{ ms}^{-1}$$

You have every reason to think that these approximations to the actual velocity at time 1 second are improving, but they are still not the actual velocity at 1 second. However, it is worthwhile making a table of average velocities over smaller and smaller time intervals.

Table 4.1 shows a table of results for the average velocity over decreasing time intervals.

Table 4.1

First value of time	Second value of time	$\dfrac{\text{increase in distance}}{\text{increase in time}}$	Average velocity
1	2	$\dfrac{5 \times 2^2 - 5 \times 1^2}{2 - 1}$	15
1	1.5	$\dfrac{5 \times 1.5^2 - 5 \times 1^2}{1.5 - 1}$	12.5
1	1.1	$\dfrac{5 \times 1.1^2 - 5 \times 1^2}{1.1 - 1}$	10.5
1	1.01	$\dfrac{5 \times 1.01^2 - 5 \times 1^2}{1.01 - 1}$	10.05
1	1.001	$\dfrac{5 \times 1.001^2 - 5 \times 1^2}{1.001 - 1}$	10.005

From the pattern of results it looks as though the average velocity is approaching 10 ms^{-1} as the time interval gets progressively smaller.

To generalize the results from this table, you could find the average velocity over a time interval δt, from $t = 1$ to $t = 1 + \delta t$.

Using the same method:

average velocity between times 1 and $1 + \delta t = \dfrac{\text{increase in distance}}{\text{increase in time}}$

$$= \frac{\delta s}{\delta t}$$
$$= \frac{5 \times (1 + \delta t)^2 - 5 \times 1^2}{(1 + \delta t) - 1}$$
$$= \frac{5 \times 2\delta t + 5\delta t^2 - 5}{\delta t}$$
$$= \frac{10\delta t + 5\delta t^2}{\delta t}$$
$$= (10 + 5\delta t) \text{ ms}^{-1}$$

You can see now, that as the time interval δt gets smaller, so the average velocity over the time interval δt gets closer to 10 ms^{-1}.

In fact, using the language of Section 2.3:

$$\text{as } \delta t \to 0, \text{ average velocity} \to 10 \text{ ms}^{-1}$$

This suggests the following definition of actual velocity which enables you to carry out calculations.

Definition:

$$\text{Velocity at time } t = \lim_{\delta t \to 0} \left(\begin{array}{c} \text{average velocity over the} \\ \text{time interval from } t \text{ to } t + \delta t \end{array} \right)$$

In the case when $s = f(t)$, you can write this as

$$\text{velocity} = \lim_{\delta t \to 0} \left(\frac{f(t + \delta t) - f(t)}{\delta t} \right)$$

Example 4.3.1

Find the velocity when $t = 2$, for the falling body for which $s = 5t^2$.

$$\text{Velocity} = \lim_{\delta t \to 0}\left(\frac{5(2+\delta t)^2 - 5\times 2^2}{\delta t}\right)$$

$$= \lim_{\delta t \to 0}\left(\frac{5(4+4\delta t+\delta t^2)-5\times 2^2}{\delta t}\right)$$

$$= \lim_{\delta t \to 0}\left(\frac{20+20\delta t+5\delta t^2-20}{\delta t}\right)$$

$$= \lim_{\delta t \to 0}\left(\frac{20\delta t+5\delta t^2}{\delta t}\right)$$

$$= \lim_{\delta t \to 0}(20+5\delta t) = 20\,\text{ms}^{-1}$$

Thus the velocity at $t = 2$ is 20 ms⁻¹.

4.4 Graphical interpretation

It is useful to look at the work of the previous section graphically. When the average velocity was calculated over the interval from 1 second to 2 seconds, you were in effect calculating the gradient of the chord PQ in the first graph in Figure 4.3. Similarly, the average velocity from 1 second to 1.5 seconds is the gradient of the chord PQ in the second graph in Figure 4.3. And similarly for the third graph. The fourth graph has the chords from the first three graphs superimposed on it, and also shows the tangent to the curve at P.

This leads to the following generalization:

▶ The velocity at time t is represented by the gradient of the tangent at the corresponding point on the displacement-time graph.

You can now modify the method developed in Section 3.4 for calculating the gradients of tangents to curves to calculating velocities.

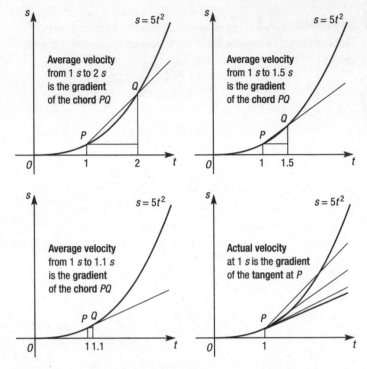

Figure 4.3

4.5 A definition of rate of change

It is now possible to give a definition of rate of change.

Definition:

The rate of change of a quantity y with respect to a quantity x at $x = a$ $= \lim\limits_{\delta x \to 0} \dfrac{f(a + \delta x) - f(a)}{\delta x}$

▶ Thus, the rate of change of a quantity y with respect to a quantity x at $x = a$ is represented by the gradient of the tangent at the point $x = a$ on the graph of y against x.

So calculating a rate of change is similar to calculating a gradient. Sometimes you simply talk about the rate of change of a quantity y without saying what the rate of change is with respect to. In that case, it is always understood that the rate of change is with respect to time. Thus, when you say that the velocity is the rate of change of displacement, it is understood to mean that velocity is the rate of change of displacement with respect to time.

Test yourself

1 The displacement in metres of a body at various times, in seconds, is given in the table below.

Time	0	1	2	3	4	5	6
Displacement	0	2	8	18	32	50	72

 a Find the average velocity over the whole period.
 b Find the average velocity over the first 3 seconds.
 c Find the average velocity over the last 3 seconds.

2 A function has the equation $y = 2x + 3$. Calculate its average rate of change over the interval
 a from $x = 1$ to $x = 2$
 b from $x = a$ to $x = b$.

3 A function is given by $f(x) = x^2 - 2x$. Find its average rate of change over the following intervals:
 a from $x = 0$ to $x = 2$
 b from $x = 0$ to $x = 1$
 c from $x = 0$ to $x = 0.1$
 d from $x = 0$ to $x = \delta x$

 What can you deduce about the actual rate of change of the function at $x = 0$?
 If the function $s = t^2 - 2t$ represents the displacement s metres of a particle in metres at time t seconds, what can you deduce about the velocity at time 0 seconds?

4 The distance in metres travelled by a body falling from rest is given approximately by the formula $s = 4.9t^2$.
 Representing an increase in time by δt and the corresponding increase in distance by δs, find (by the method of Section 4.3) an expression for δs in terms of δt for $t = T$.
 Hence find the value of $\frac{\delta s}{\delta t}$ for $t = T$.

Use this result to find the average velocity for the following intervals:

a 2 s to 2.2 s
b 2 s to 2.1 s
c 2 s to 2.01 s
d 2 s to 2.001 s

From these results, deduce the velocity at 2 seconds.

Key ideas

▶ As you saw in Chapter 3, if you consider a point, P, on a curve, (x, y) and also a point very close to it $(x + \delta x, y + \delta y)$, the gradient of the chord joining the points is $\frac{\delta y}{\delta x}$. As the second point gets closer and closer to P, the values of δx and δy both tend to zero, but their ratio tends to the value of the gradient of the curve at P.

▶ As δx and δy both tend to zero, their ratio is written as $\frac{\delta y}{\delta x}$. This is the differential coefficient of the function.

▶ The differential coefficient of a constant is zero.

▶ The differential coefficient, with respect to x, of mx is m.

The differential coefficient, with respect to x, of ax^n is nax^{n-1}.

5

Differentiation

In this chapter you will learn:

▶ *how to differentiate simple functions*
▶ *some of the notation used for differentiation*
▶ *the meaning of the sign of the derivative.*

5.1 Algebraic approach to the rate of change of a function

The work of this chapter takes an important step forward in systematizing the work of the previous two chapters. Here is a brief summary of the steps of Chapter 3 and Chapter 4.

▶ The value of a function changes as the variable on which it depends changes.

▶ The gradient of a curve at a point $x = a$ is calculated using a limit process (shown in Section 3.4).

▶ The average rate of change of a function y with respect to x over an interval is geometrically equivalent to the gradient of the chord on the graph of y against x over the same interval.

▶ The rate of change of a function y with respect to x at $x = a$ is geometrically equivalent to the gradient of the tangent at $x = a$ on the graph of y against x.

In Example 3.4.3, you saw that the gradient at $x = a$ of the graph of $y = x^2$ was $2a$. For convenience, this work is repeated here, but in a slightly different form.

$$
\begin{aligned}
\text{Gradient} &= \lim_{\delta x \to 0} \left(\frac{(a + \delta x)^2 - a^2}{\delta x} \right) \\
&= \lim_{\delta x \to 0} \left(\frac{(a^2 + 2a\delta x + \delta x^2) - a^2}{\delta x} \right) \\
&= \lim_{\delta x \to 0} \left(\frac{a^2 + 2a\delta x + \delta x^2 - a^2}{\delta x} \right) \\
&= \lim_{\delta x \to 0} \left(\frac{2a\delta x + \delta x^2}{\delta x} \right) \\
&= \lim_{\delta x \to 0} \left(2a + \delta x \right) = 2a
\end{aligned}
$$

This shows that the gradient depends on the value of x, and is therefore a function of x. The definition in Section 4.5 also shows that the rate of change of the function $f(x) = x^2$ is $2a$

when $x = a$. You can summarize the process above geometrically, by saying that at a given point x you find the increase in y (called δy) for a given increase in x (called δx) using the fact that $\delta y = f(x + \delta x) - f(x)$. You then use the definitions in Section 3.5 to say that:

$$\left.\begin{array}{r}\text{gradient of}\\ f(x)\ \text{at}\ a\end{array}\right\} = \lim_{\delta x \to 0}\left(\frac{\delta y}{\delta x}\right) = \lim_{\delta x \to 0}\left(\frac{f(a + \delta x) - f(a)}{\delta x}\right)$$

When calculating the gradient of a function at a point, it is more usual simply to call the point x than to call it $x = a$. Thus the previous remark becomes

$$\left.\begin{array}{r}\text{gradient of}\\ f(x)\ \text{at}\ x\end{array}\right\} = \lim_{dx \to 0}\left(\frac{\delta y}{\delta x}\right) = \lim_{\delta x \to 0}\left(\frac{f(x + \delta x) - f(x)}{\delta x}\right)$$

Example 5.1.1

Find the gradient of $y = 2x^2$

Let $y = 2x^2$

Then $y + \delta y = 2(x + \delta x)^2$

So, by subtracting,
$$\begin{aligned}\delta y &= 2(x + \delta x)^2 - 2x^2\\ &= 2x^2 + 4x\,\delta x + 2(\delta x)^2 - 2x^2\\ &= 4x\,\delta x + 2(\delta x)^2\end{aligned}$$

Dividing by δx gives
$$\frac{\delta y}{\delta x} = 4x + 2\delta x$$

To find the gradient, you now have to find the limit of $\frac{\delta y}{\delta x}$ as $\delta x \to 0$.

Therefore $\left.\begin{array}{r}\text{gradient of}\\ y = 2x^2\ \text{at}\ x\end{array}\right\} = \lim_{\delta x \to 0}\left(\frac{\delta y}{\delta x}\right) = \lim_{\delta x \to 0}(4x + 2\delta x) = 4x$

Now, for example, you can calculate the gradient at $x = 1$ by substituting $x = 1$; and you find that the gradient is 4. Similarly, the gradient when $x = -2$ is -8.

Example 5.1.2

Find the gradient of $y = f(x)$ where $f(x) = x^2 + 2x$

Let $f(x) = x^2 + 2x$

Then $f(x + \delta x) = (x + \delta x)^2 + 2(x + \delta x)$

So, by subtracting,

$$
\begin{aligned}
f(x + \delta x) - f(x) &= (x + \delta x)^2 + 2(x + \delta x) - x^2 - 2x \\
&= x^2 + 2x\delta x + (\delta x)^2 + 2x + 2\delta x - x^2 - 2x \\
&= 2x\delta x + (\delta x)^2 + 2\delta x
\end{aligned}
$$

Dividing by δx, gives
$$
\frac{f(x + \delta x) - f(x)}{\delta x} = 2x + \delta x + 2
$$

Therefore

$$
\left.\begin{array}{l} \text{gradient of} \\ y = x^2 + 2x \text{ at } x \end{array}\right\} = \lim_{\delta x \to 0} \frac{f(x + \delta x) - f(x)}{\delta x}
$$

$$
= \lim_{\delta x \to 0} (2x + \delta x + 2) = 2x + 2
$$

Using this result, you can calculate the gradient at the point $x = 3$ by substituting $x = 3$ in the expression for gradient $2x + 2$; you find that the gradient is 8.

The notation $\frac{dy}{dx}$ is used to denote $\lim_{\delta x \to 0} \frac{\delta y}{\delta x}$, in which the English letter d is used instead of the Greek letter δ, and the condition $\delta x \to 0$ is understood. Thus in Example 5.1.2 above, you would write at the end,

$$
\frac{dy}{dx} = \lim_{\delta x \to 0} \frac{\delta y}{\delta x} = \lim_{\delta x \to 0} (2x + \delta x + 2) = 2x + 2
$$

Definition: The expression $\frac{dy}{dx}$ is called the **derivative** or the **derived function** of the function y with respect to x, the independent variable.

Thus, when $y = x^2 + 2x$, the derivative of y with respect to x is $2x + 2$; or you could say that the derivative of $x^2 + 2x$ with respect to x is $2x + 2$. You can use similar methods to those of Examples 5.1.1 and 5.1.2 to find the derivative of any other function.

5.2 The derived function

Here is a summary of Section 5.1.

▶ If y is a continuous function of x and δx is an increase in the value of x, there will be a corresponding increase (or decrease) of δy in y.

▶ The ratio $\frac{\delta y}{\delta x}$ represents the average rate of increase of y with respect to x, when x is increased from x to $x + \delta x$.

▶ Since y is a continuous function of x, if δx becomes indefinitely small, so also does δy.

▶ When $\delta x \to 0$, the ratio $\frac{\delta y}{\delta x}$ in general tends to a finite limit. This limit is called the derivative of y with respect to x, and is denoted by the symbol $\frac{dy}{dx}$. So $\frac{dy}{dx} = \lim\limits_{\delta x \to 0} \frac{\delta y}{\delta x}$

The differential calculus is concerned with the variation of functions. You can think of the derivative as a rate-measurer in these variations. Thus, for the function $y = x^2$, $\frac{dy}{dx} = 2x$. When $x = 4$, y (that is, x^2) is changing its value at eight times the rate at which x is changing.

The derivative $\frac{dy}{dx}$ is also called the **differential coefficient** of y with respect to x.

Nugget – Does it work with a new letter?

You have just seen that, if $y = x^2$, then $\frac{dy}{dx} = 2x$.

It is important to remember that $\frac{dy}{dx}$ means 'the differential of y with respect to x'. In other words, this result only works for differentiating terms in x.

Suppose you are given a similar statement but with a different letter as follows: $y = a^2$. You cannot write that $\frac{dy}{dx} = 2a$ because you have differentiated with respect to x, not a as was required. It would, however, be correct to write: $\frac{dy}{da} = 2a$

So a useful tip to avoid this error is always to check that the letters match up.

5.3 Notation for the derivative

Besides the form $\frac{dy}{dx}$, the derivative of y with respect to x may also be denoted by y' (read as y prime).

Thus if $y = x^2$, $y' = 2x$.

If you are using function notation, the derivative of $f(x)$ is denoted by $f'(x)$. Thus, in function notation, the general **definition of the derivative** $f'(x)$ of any function $f(x)$ is given by

$$f'(x) = \lim_{\delta x \to 0} \left(\frac{f(x + \delta x) - f(x)}{\delta x} \right)$$

This definition is a restatement of the definition in Section 5.1.

The same forms of notation are used when other letters are used to represent functions. Thus if s is a function of t, the derivative of s with respect to t is written $\frac{ds}{dt}$. Or if $s = f(t)$, $\frac{ds}{dt} = f'(t)$. The process of finding the derivative or the differential coefficient of a function is called **differentiation**. You can express the operation of differentiation by using the **operating symbol** $\frac{d}{dx}$. Thus differentiating x^2 with respect to x can be written in the form $\frac{d(x^2)}{dx}$ or $\frac{d}{dx}(x^2)$.

In general, differentiating $f(x)$ with respect to x can be denoted by

$$\frac{d(f(x))}{dx} \quad \text{or} \quad \frac{d}{dx}(f(x))$$

Another notation sometimes used for differentiating the function y with respect to x is $D_x y$. If there is no possible confusion about what the independent variable might be, the notation Dy is used.

5.4 Differentials

The form $\frac{dy}{dx}$, which appears to be a quotient, suggests that there are quantities dy and dx which divide to make it. It is possible to interpret dy and dx in this way.

Figure 5.1 shows a curve $y = f(x)$ together with a tangent at a point P. The distance PR shows an increase which you can think of as either δx or as dx. The distance RQ, which is the corresponding distance from R to the curve, is called δy. The distance from R to where the tangent line meets RQ is called dy.

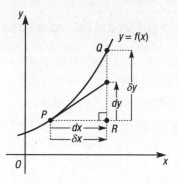

Figure 5.1

The distances dy and dx are called **differentials**. The differential coefficient can be regarded as the ratio of these differentials. For the example $y = x^2$ in which $\frac{dy}{dx} = 2x$, you can write

$$dy = 2x\,dx$$

In this form $2x$ is the coefficient of the differential dx. This explains the term 'differential coefficient'. At this stage there is no advantage in thinking of the differential coefficient in this way.

5.5 Sign of the derivative

The derivative of a function is equal to the gradient of the function. It was shown in Section 3.6 that the gradient may be positive or negative, so it follows that the derivative can be positive or negative.

Similarly, if the derivative is positive, the function is increasing, and if the derivative is negative, the function is decreasing.

5.6 Some examples of differentiation

In Examples 5.6.1 to 5.6.6 a number of functions are differentiated from first principles, that is, by going back to the definition of differentiation in Section 5.2. However, in Examples 5.6.5 and 5.6.6 the δy and δx notation is used. It is important that you realize that the mathematics lying behind the two notations is equivalent. If for $f(x)$ you read y, for $f(x + \delta x)$ you read $y + \delta y$, and for $f(x + \delta x) - f(x)$ you read δy, the two treatments become identical.

Example 5.6.1 Differentiating a constant

Since a derivative measures the gradient of a graph of a function, and the graph of a constant is horizontal and therefore has gradient zero, the derivative of a constant is zero.

Alternatively, let $f(x) = c$, where c is a constant.

Using the definition of derivative given in Section 5.4,

$$f'(x) = \lim_{\delta x \to 0} \left(\frac{f(x + \delta x) - f(x)}{\delta x} \right)$$

$$= \lim_{\delta x \to 0} \left(\frac{c - c}{\delta x} \right)$$

$$= \lim_{\delta x \to 0} \left(\frac{0}{\delta x} \right)$$

$$= 0$$

Thus the derivative of a constant is zero.

Example 5.6.2 Differentiating $f(x) = mx + c$

In Section 3.2 you saw that $y = mx + c$ is the equation of a straight line of gradient m. Therefore if $f(x) = mx + c$, then $f'(x) = m$.

Alternatively, let $f(x) = mx + c$ where m and c are constants. Then

$$f'(x) = \lim_{\delta x \to 0} \left(\frac{f(x + \delta x) - f(x)}{\delta x} \right)$$

$$= \lim_{\delta x \to 0} \left(\frac{m(x + \delta x) + c - (mx + c)}{\delta x} \right)$$

$$= \lim_{\delta x \to 0} \left(\frac{(mx + m\delta x + c - mx - c)}{\delta x} \right)$$

$$= \lim_{\delta x \to 0} \left(\frac{m\delta x}{\delta x} \right)$$

$$= \lim_{\delta x \to 0} m$$

$$= m$$

Notice that the gradient $f'(x)$ depends only on the value of m and not on c. So the equation $y = mx + c$, which leads to $\frac{dy}{dx} = m$, represents a set of parallel lines, each having the gradient m. See Section 3.2.

Example 5.6.3 Differentiating $f(x) = x^3$

Here is another example of the general method for finding a derivative from first principles, where $f(x) = x^3$.

$$f'(x) = \lim_{\delta x \to 0} \left(\frac{f(x + \delta x) - f(x)}{\delta x} \right)$$

$$= \lim_{\delta x \to 0} \left(\frac{(x + \delta x)^3 \, \delta x^3}{\delta x} \right)$$

$$= \lim_{\delta x \to 0} \left(\frac{x^3 + 3x^2 \delta x + 3x(\delta x)^2 + (\delta x)^3 - x^3}{\delta x} \right)$$

$$= \lim_{\delta x \to 0} \left(\frac{3x^2 \delta x + 3x(\delta x)^2 + (\delta x)^3}{\delta x} \right)$$

$$= \lim_{\delta x \to 0} \left(3x^2 + 3x \delta x + (\delta x)^2 \right)$$

$$= 3x^2$$

Example 5.6.4 Differentiating $f(x) = x^n$

To differentiate $f(x) = x^4$ using the method of Example 5.6.3, you need the expansion of $(x + \delta x)^4$, that is,

$$(x + \delta x)^4 = x^4 + 4x^3 \, \delta x + 6x^2(\delta x)^2 + 4x(\delta x)^3 + (\delta x)^4$$

Then

$$f'(x) = \lim_{\delta x \to 0} \left(\frac{f(x + \delta x) - f(x)}{\delta x} \right)$$

$$= \lim_{\delta x \to 0} \left(\frac{(x + \delta x)^4 - x^4}{\delta x} \right)$$

$$= \lim_{\delta x \to 0} \left(\frac{x^4 + 4x^3 \delta x + 6x^2(\delta x)^2 + 4x(\delta x)^3 + (\delta x)^4 - x^4}{\delta x} \right)$$

$$= \lim_{\delta x \to 0} \left(\frac{4x^3 \delta x + 6x^2(\delta x)^2 + 4x(\delta x)^3 + (\delta x)^4}{\delta x} \right)$$

$$= \lim_{\delta x \to 0} \left(4x^3 + 6x^2 \delta x + 4x(\delta x)^2 + (\delta x)^3 \right)$$

$$= 4x^3$$

You can deal with the function $f(x) = x^n$, where n is a positive integer, in a similar way by using the binomial theorem to expand the product $(x + \delta x)^n$. In this expansion, it is the second term which gives rise to the derivative.

Thus, for example, if $f(x) = x^5$, then $f'(x) = 5x^4$

Similarly, if $f(x) = x^6$, then $f'(x) = 6x^5$

Generally for a function, $f(x) = x^n$ where n is a positive integer, you can deduce that $f'(x) = nx^{n-1}$

Here is a proof using the binomial theorem.

First note that:

$$(x + \delta x)^n = x^n + nx^{n-1}\delta x + \frac{n(n-1)}{2!}x^{n-2}(\delta x)^2 + \cdots$$

so $$(x + \delta x)^n - x^n = nx^{n-1}\delta x + \frac{n(n-1)}{2!}x^{n-2}(\delta x)^2 + \cdots$$

Therefore

$$f'(x) = \lim_{\delta x \to 0}\left(\frac{f(x + \delta x) - f(x)}{\delta x}\right)$$

$$= \lim_{\delta x \to 0}\left(\frac{nx^{n-1}\delta x + \frac{n(n-1)}{2!}x^{n-2}(\delta x)^2 + \cdots}{\delta x}\right)$$

$$= \lim_{\delta x \to 0}\left(nx^{n-1} + \frac{n(n-1)}{2!}x^{n-2}\delta x + \cdots\right)$$

$$= nx^{n-1}$$

The question now arises, does this result hold for values of n other than positive integers? The answer is yes. The result that if $f(x) = x^n$ then $f'(x) = nx^{n-1}$ holds for all real values of n, positive negative and zero, but this cannot be proved by using the binomial theorem.

The conclusion is therefore, that if $f(x) = x^n$, then $f'(x) = nx^{n-1}$, or alternatively $\frac{d}{dx}(x^n) = nx^{n-1}$ **for all values of n**.

Example 5.6.5 Differentiating $y = ax^n$, where a is a constant

In this example and the next, $\frac{dy}{dx}$ notation is used.

If $y = ax^n$, then $y + \delta y = a(x + \delta x)^n$, so:

$$y + \delta y = a\left(x^n + nx^{n-1}\delta x + \frac{n(n-1)}{2!}x^{n-2}(\delta x)^2 + \cdots \right)$$

Therefore $\delta y = a\left(nx^{n-1}\delta x + \frac{n(n-1)}{2!}x^{n-2}(\delta x)^2 + \cdots \right)$

and $\qquad \dfrac{\delta y}{\delta x} = a\left(nx^{n-1} + \frac{n(n-1)}{2!}x^{n-2}\,\delta x + \cdots \right)$

So $\dfrac{dy}{dx} = \lim\limits_{\delta x \to 0}\dfrac{\delta y}{\delta x} = \lim\limits_{\delta x \to 0}\left(a\left(nx^{n-1} + \frac{n(n-1)}{2!}x^{n-2}\delta x + \cdots \right) \right) = anx^{n-1}$

Nugget – A rule for differentiating x^n

The result given above is a very useful one. One way of remembering it is with the rule: **multiply by the power, reduce the power by one**.

For example, let's try differentiating $y = 2x^5$

▶ *multiplying* by the power means multiplying by 5

▶ *reducing* the power by one means that the 5 becomes 4.

So the answer is $5 \times 2x^4 = 10x^4$

Example 5.6.6 Differentiating $y = \frac{1}{x}$

If $y = \frac{1}{x}$ then $y + \delta y = \frac{1}{x + \delta x}$

Therefore $\delta y = \dfrac{1}{x + \delta x} - \dfrac{1}{x} = \dfrac{x - (x + \delta x)}{x(x + \delta x)} = \dfrac{-\delta x}{x(x + \delta x)}$

so $\qquad \dfrac{\delta y}{\delta x} = \dfrac{-1}{x(x + \delta x)}$

Therefore
$$\frac{dy}{dx} = \lim_{\delta x \to 0} \frac{\delta y}{\delta x} = \lim_{\delta x \to 0} \frac{-1}{x(x + \delta x)} = \frac{-1}{x^2}$$

Notice that the result, if $y = \frac{1}{x} = x^{-1}$ then $\frac{dy}{dx} = -\frac{1}{x^2} = -x^{-2}$ fits

the general rule that $\frac{d}{dx}(x^n) = nx^{n-1}$

Example 5.6.7

Write down the derivatives of the following functions.

1 $y = x^8$ $\qquad$ $\frac{dy}{dx} = 8x^{8-1} = 8x^7$

2 $y = x^{\frac{1}{2}}$ $\qquad$ $\frac{dy}{dx} = \frac{1}{2}x^{\frac{1}{2}-1} = \frac{1}{2}x^{-\frac{1}{2}} = \frac{1}{2\sqrt{x}}$

3 $y = x^{-3}$ $\qquad$ $\frac{dy}{dx} = -3x^{-3-1} = -3x^{-4} = \frac{-3}{x^4}$

4 $y = x^{1.5}$ $\qquad$ $\frac{dy}{dx} = 1.5x^{1.5-1} = 1.5x^{0.5} = 1.5\sqrt{x}$

5 $y = x^{\frac{-1}{3}}$ $\qquad$ $\frac{dy}{dx} = \left(-\frac{1}{3}\right)x^{-\frac{1}{3}-1} = \left(-\frac{1}{3}\right)x^{-\frac{4}{3}}$

6 $y = x$ $\qquad$ $\frac{dy}{dx} = 1x^{1-1} = 1x^0 = 1$

Example 5.6.8

Differentiate the following functions.

1 $y = 6x^4$ $\qquad$ $\frac{dy}{dx} = 6 \times 4 \times x^{4-1} = 24x^3$

2 $y = 4\sqrt[3]{x}$ then $y = 4x^{\frac{1}{3}}$ and

$$\frac{dy}{dx} = 4 \times \frac{1}{3} \times x^{\frac{1}{3}-1} = \frac{4}{3}x^{-\frac{2}{3}} = \frac{4}{3x^{\frac{2}{3}}} = \frac{4}{3\sqrt[3]{x^2}}$$

3 $y = px^{2q}$ $\qquad$ $\frac{dy}{dx} = 2q \times p \times x^{2q-1} = 2pqx^{2q-1}$

4 $s = 16t^2$ $\qquad$ $\frac{ds}{dt} = 2 \times 16 \times t^{2-1} = 32t$

Example 5.6.9

Find the gradient of the tangent to the curve $y = \dfrac{1}{x}$ at the point where $x = 1$.

The gradient is given by the value of $\dfrac{dy}{dx}$ when $x = 1$

Now $\dfrac{d}{dx}\left(\dfrac{1}{x}\right) = \dfrac{d}{dx}\left(x^{-1}\right) - x^{-2} = -\dfrac{1}{x^2}$

When $x = 1$, $\dfrac{dy}{dx} = -1$, so the gradient is -1

Test yourself

1. Write down the derivatives of the following functions with respect to x.

x^7 $5x$ $\dfrac{x}{3}$ $0.06x$ $\dfrac{1}{4}x^5$ $15x^4$ $\dfrac{2x^6}{3}$ $1.5x^3$ $(4x)^2$

2. Differentiate the following with respect to x.

bx^4 $\dfrac{ax^6}{b}$ ax^p x^{2a} $2x^{2b+1}$ $4\pi x^2$

3. Differentiate the following with respect to x.

$6x + 4$ $0.54x - 6$ $-3x + 2$ $px + q$

4. Of what functions of x are the following the differential coefficients?

x $3x$ x^2 $\tfrac{1}{4}x^2$ x^5 x^n x^{2a} $\tfrac{2}{3}x^3$ $4ax^2$

5. Let $v = u + at$, where u and a are constants. Find $\dfrac{dv}{dt}$

6. Let $s = \tfrac{1}{2}at^2$, where a is a constant. Find $\dfrac{ds}{dt}$ when $a = 20$

7. Let $A = \pi r^2$. Find $\dfrac{dA}{dr}$

8. Let $V = \tfrac{4}{3}\pi r^3$. Find $\dfrac{dV}{dr}$

9. Differentiate the following functions of x.

$$5\sqrt{x} \qquad \frac{5}{x} \qquad \frac{5}{\sqrt{x}} \qquad \sqrt[3]{x^2} \qquad \sqrt[4]{2x^3}$$

10. Differentiate the following with respect to x.

$$x^{0.4} \qquad 8x^{0.2} \qquad \frac{8}{x^{0.2}} \qquad 6x^{-4} \qquad x^{-p}$$

11. Differentiate the following with respect to x.

$$6x^{3.2} \qquad 2x^{-1.5} \qquad 29x^{0.7} \qquad \frac{6}{\sqrt[5]{x^3}}$$

12. Let $p = \dfrac{20}{v^2}$. Find $\dfrac{dp}{dv}$

13. Find the gradient of the curve $y = \frac{1}{4}x^2$ at the point where $x = 3$. For what value of x is the gradient of the curve equal to zero?

14. Find the gradient of the curve $y = 2x^3$ at the point where $x = 2$.

15. Find the gradients of the curve $y = \dfrac{2}{x}$ at the points where $x = 10, 2, 1, \frac{1}{2}$.

16. Find from first principles the differential coefficient of $y = \dfrac{1}{x^2}$

17. At what point on the graph of x^2 is the gradient of the curve equal to 2?

18. At what points on the graph of $y = x^3$ does the tangent to the curve make an angle of 45° with the x-axis?

19. At what point on the graph of $y = \sqrt{x}$ is the gradient equal to 2?

20. It is required to draw a tangent to the curve $y = 0.5x^2$ which is parallel to the straight line $2x - 4y = 3$. At what point on the curve should it be drawn?

6

Some rules for differentiation

In this chapter you will learn:

▶ *how to differentiate more complicated functions*
▶ *how to differentiate a function of a function*
▶ *how to differentiate implicit functions.*

6.1 Differentiating a sum

In Chapter 5 the functions that were differentiated in Examples 5.6.1 and 5.6.3 to 5.6.6 all had just one term; the function in Example 5.6.2, $f(x) = mx + c$, or $y = mx + c$ had two terms, but one is a constant.

Here is a general treatment for differentiating a function which is itself the sum of two or more functions of the same variable, such as $y = 5x^3 + 14x^2 - 7x$. The proof given below is a general one for the sum of any number of functions of the same variable.

Let u and v be functions of x, and let y be their sum so that $y = u + v$.

Let x be increased by δx.

Then u, v and y, being functions of x, will increase correspondingly.

Call these increases δu, δv and δy, so that:

$$u \text{ becomes } u + \delta u$$
$$v \text{ becomes } v + \delta v$$
$$y \text{ becomes } y + \delta y$$

From $\qquad\qquad\qquad y = u + v$

you find $\qquad\qquad y + \delta y = (u + \delta u) + (v + \delta v)$

By subtraction $\qquad \delta y = \delta u + \delta v$

Dividing by δx $\qquad \dfrac{\delta y}{\delta x} = \dfrac{\delta u}{\delta x} + \dfrac{\delta v}{\delta x}$

This is true for all values of δx and the corresponding increases δu, δv and δy.

Therefore the limit of $\dfrac{\delta y}{\delta x}$ is equal to the limit of $\dfrac{\delta u}{\delta x} + \dfrac{\delta v}{\delta x}$ as $\delta x \to 0$ (by Theorem 1 in Section 2.6).

Therefore

$$\lim_{\delta x \to 0} \frac{\delta y}{\delta x} = \lim_{\delta x \to 0} \left(\frac{\delta u}{\delta x} + \frac{\delta v}{\delta x} \right)$$

$$= \lim_{\delta x \to 0} \left(\frac{\delta u}{\delta x} \right) + \lim_{\delta x \to 0} \left(\frac{\delta v}{\delta x} \right) \text{ (using Theorem 2 of Section 2.6)}$$

Therefore, by replacing these forms by the corresponding expressions for their derivatives:

$$\frac{dy}{dx} = \frac{du}{dx} + \frac{dv}{dx}$$

This is the rule for differentiating a sum. It holds for the sum of any number of functions.

▶ The derivative of the sum of a number of functions is equal to the sum of the derivatives of these functions.

Nugget – A rule for differentiating a sum or difference

So, as you can see, differentiating a sum or difference is easy – you just treat each term separately!

Example 6.1.1

Differentiate $y = 3x^3 + 7x^2 - 9x + 20$ with respect to x.

Using the rule for sums $\frac{dy}{dx} = 9x^2 + 14x - 9$

Example 6.1.2

Find the gradient of the graph of $y = x^2 - 4x + 3$ at $x = 3$. Where is the point of zero gradient on this graph?

If $y = x^2 - 4x + 3$, then $\frac{dy}{dx} = 2x - 4$. When $x = 3$, $\frac{dy}{dx} = 2 \times 3 - 4 = 2$

When the gradient is zero, $\frac{dy}{dx} = 0$ so $2x - 4 = 0$. Therefore $x = 2$

When $x = 2$, $y = 2^2 - 4 \times 2 + 3 = -1$

The required point has coordinates $(2, -1)$

Example 6.1.3

Let $s = 80t - 16t^2$. Find $\frac{ds}{dt}$, and find t when $\frac{ds}{dt} = 16$.

If $s = 80t - 16t^2$, $\frac{ds}{dt} = 80 - 32t$

When $\frac{ds}{dt} = 16$, $80 - 32t = 16$. Therefore $32t = 64$, and $t = 2$

Exercise 6.1

1 Differentiate the following functions with respect to x.

  a $6x^2 + 5x$ **b** $3x^3 + x - 1$

  c $4x^4 + 3x^2 - x$ **d** $\frac{1}{2}x^2 + \frac{1}{7}x + \frac{1}{4}$

  e $\frac{5}{x} + 4x$ **f** $7 + \frac{4}{x} - \frac{2}{x^2}$

  g $x(5 - x + 3x^2)$ **h** $8\sqrt{x} + \sqrt{10}$

2 Find $\frac{ds}{dt}$ for each of the following.

  a $s = ut + \frac{1}{2}at^2$, where u and a are constants

  b $s = 5t + 16t^2$

  c $s = 3t^2 - 4t + 7$

3 Find $\frac{dy}{dx}$ when $y = ax^3 + bx^2 + cx + d$, where a, b, c and d are constants.

4 Differentiate $\left(x + \frac{1}{x}\right)^2$ with respect to x.

5 Differentiate $\sqrt{x} + \frac{1}{\sqrt{x}}$ with respect to x.

6 Differentiate $(1 + x)^3$ with respect to x.

7 If $y = x^{2n} - nx^2 + 5n$, where n is constant, find $\frac{dy}{dx}$.

8 Find $\frac{dy}{dx}$ when $y = \sqrt{x} + \sqrt[3]{x} + \frac{2}{x}$.

9 Find the gradient at $x = 1.5$ on the graph of $y = 2x^2 - 3x + 1$. For what value of x will the gradient be zero?

10 For what values of x will the graph of $y = x(x^2 - 12)$ have zero gradient?

11 What are the gradients of the curve $y = x^3 - 6x^2 + 11x - 6$ when x has the values 1, 2 and 3?

12 Find the coordinates of the points of zero gradient on the graph $y = x + \frac{1}{x}$.

6.2 Differentiating a product

You can find the derivative of some products such as $(x + 2)^3$ or $3x(x + 2)$ by multiplying out the brackets and using the rule for finding the derivative of a sum. In many cases, however, you cannot do that, as, for example, in $x^2\sqrt{1 - x}$ and $x^3 \sin x$.

The derivative of a product is not equal to the product of the derivatives, as you can verify by testing an example such as $3x(x + 2)$.

Here is a method for finding a general rule.

Let u and v be functions of x, and let y be their product so that $y = u \times v$, which is also a function of x.

Let x be increased by δx.

Then u, v and y, being functions of x, will receive corresponding increases.

Call these increases δu, δv and δy, so that

$$u \text{ becomes } u + \delta u$$
$$v \text{ becomes } v + \delta v$$
$$y \text{ becomes } y + \delta y$$

Then $y + \delta y = (u + \delta u)(v + \delta v)$ and $y = uv$ so that:

$$\delta y = (u + \delta u)(v + \delta v) - uv$$
$$= uv + u(\delta v) + v(\delta u) + (\delta u)(\delta v) - uv$$
$$= u(\delta v) + v(\delta u) + (\delta u)(\delta v)$$

Dividing by δx: $\qquad \dfrac{\delta y}{\delta x} = u\dfrac{\delta v}{\delta x} + v\dfrac{\delta u}{\delta x} + \delta u\dfrac{\delta v}{\delta x}$

Now let $\delta x \to 0$ Then $\delta u \to 0$.

Then, using Theorems 1 and 2 from Section 2.6:

$$\lim_{\delta x \to 0} \frac{\delta y}{\delta x} = \lim_{\delta x \to 0}\left(u\frac{\delta v}{\delta x} + v\frac{\delta u}{\delta x} + \delta u\frac{\delta v}{\delta x} \right)$$
$$= \lim_{\delta x \to 0}\left(u\frac{\delta v}{\delta x} \right) + \lim_{\delta x \to 0}\left(v\frac{\delta u}{\delta x} \right) + \lim_{\delta x \to 0}\left(\delta u\frac{\delta v}{\delta x} \right)$$

In the last term on the right-hand side, you can use Theorem 3 from Section 2.6, and get:

$$\lim_{\delta x \to 0} \frac{\delta y}{\delta x} = \lim_{\delta x \to 0}\left(u\frac{\delta v}{\delta x} \right) + \lim_{\delta x \to 0}\left(v\frac{\delta u}{\delta x} \right) + \lim_{\delta x \to 0}(\delta u)\lim_{\delta x \to 0}\left(\frac{\delta v}{\delta x} \right)$$

Replacing these forms by the corresponding expressions for their derivatives, and recalling that $\delta u \to 0$ as $\delta x \to 0$:

$$\frac{dy}{dx} = u\frac{dv}{dx} + v\frac{du}{dx} + 0 \times \frac{dv}{dx}$$

Therefore
$$\frac{dy}{dx} = u\frac{dv}{dx} + v\frac{du}{dx}$$

▶ The derivative of a product of two factors is the sum of the first factor multiplied by the derivative of the second and the second factor multiplied by the derivative of the first.

Nugget – A rule for differentiating a product involving two factors

The summary of this rule can be condensed even further as:

Leave the first, times differentiate the second, plus leave the second times differentiate the first. This is illustrated below.

$$y = uv \qquad \frac{dy}{dx} = u\frac{dv}{dx} + v\frac{du}{dx}$$

I suggest you read this rule through several times and then, when you think you understand it, see if you can apply it in the worked examples on the following page.

You can extend this rule to a product of more than two factors. Thus if $y = uvw$, you can write this as $y = (uv)w$. Then:

$$\frac{dy}{dx} = w\frac{d}{dx}(uv) + uv\frac{dw}{dx} = w\left(u\frac{dv}{dx} + v\frac{du}{dx}\right) + uv\frac{dw}{dx}$$

$$= vw\frac{du}{dx} + wu\frac{dv}{dx} + uv\frac{dw}{dx}$$

Example 6.2.1

Differentiate $(x^2 - 5x + 2)(2x^2 + 7)$

Let $y = (x^2 - 5x + 2)(2x^2 + 7)$

Then

$$\frac{dy}{dx} = \left\{\frac{d(x^2 - 5x + 2)}{dx} \times (2x^2 + 7)\right\} + \left\{\frac{d(2x^2 + 7)}{dx} \times (x^2 - 5x + 2)\right\}$$

$$= (2x - 5)(2x^2 + 7) + 4x(x^2 - 5x + 2)$$

You can simplify this further, if you wish, by multiplying out the brackets and collecting the like terms. In that case you find that:

$$\frac{dy}{dx} = 8x^3 - 30x^2 + 22x - 35$$

Example 6.2.2

Differentiate $y = (x^2 - 1)(2x + 1)(x^3 + 2x^2 + 1)$

Using the extension of the product rule:

$$\frac{d}{dx}\left\{(x^2 - 1)(2x + 1)(x^3 + 2x^2 + 1)\right\}$$

$$= \left\{\frac{d(x^2 - 1)}{dx} \times (2x + 1)(x^3 + 2x^2 + 1)\right\}$$

$$+ \left\{\frac{d(2x + 1)}{dx} \times (x^2 - 1)(x^3 + 2x^2 + 1)\right\}$$

$$+ \left\{\frac{d(x^3 + 2x^2 + 1)}{dx} \times (x^2 - 1)(2x + 1)\right\}$$

$$= 2x(2x + 1)(x^3 + 2x^2 + 1) + 2(x^2 - 1)(x^3 + 2x^2 + 1)$$

$$+ (3x^2 + 4x)(x^2 - 1)(2x - 1)$$

You can simplify this further, if you wish, to get

$$12x^5 + 25x^4 - 9x^2 - 2x - 2$$

Exercise 6.2

Differentiate the following using the rule for products.

1 $(3x + 1)(2x + 1)$ **2** $(x^2 + 1)(\frac{1}{2}x + 1)$

3 $(3x - 5)(x^2 + 2x)$ **4** $(x^2 + 3)(2x^2 - 1)$

5 $(x^2 + 4x)(3x^2 - x)$ **6** $(x^2 + x + 1)(x - 1)$

7 $(x^2 - x + 1)(x + 1)$ **8** $(x^2 + 4x + 5)(x^2 - 2)$

9 $(x^2 - 5)(x^2 + 5)$ **10** $(x^2 - x + 1)(x^2 + x - 1)$

11 $(x - 2)(x^2 + 2x + 4)$ **12** $(2x^2 - 3)(3x^2 + x - 1)$

13 $(x - 1)(2x + 1)(3x + 2)$ **14** $(ax^2 + bx + c)(px + q)$

15 $2x^{\frac{3}{2}}\left(\sqrt{x} + 2\right)\left(\sqrt{x} - 1\right)$

6.3 Differentiating a quotient

In Example 5.6.6 the derivative of a simple quotient, namely $\frac{1}{x}$ was derived from first principles. This method, however, becomes very tedious. Here is a derivation of the general rule. Let u and v be functions of x, and let y be their quotient so that $y = \frac{u}{v}$ which is also a function of x. Using the notation of Section 6.2:

$$y = \frac{u}{v}$$

and

$$y + \delta y = \frac{u + \delta u}{v + \delta v}$$

Subtracting

$$\delta y = \frac{u + \delta u}{v + \delta v} - \frac{u}{v}$$

$$= \frac{v(u + \delta u) - u(v + \delta v)}{v(v + \delta v)}$$

$$= \frac{v\delta u - u\delta v}{v(v + \delta v)}$$

Dividing by δx

$$\frac{\delta y}{\delta x} = \frac{v\dfrac{\delta u}{\delta x} - u\dfrac{\delta v}{\delta x}}{v(v + \delta v)}$$

Let $\delta x \to 0$. Then δu, δv and δy tend to zero, and, using the limit theorems in Section 2.6:

$$\lim_{\delta x \to 0}\left(\frac{\delta y}{\delta x}\right) = \frac{\displaystyle\lim_{\delta x \to 0}\left(v\frac{\delta u}{\delta x}\right) - \lim_{\delta x \to 0}\left(u\frac{\delta v}{\delta x}\right)}{\displaystyle\lim_{\delta x \to 0} v(v + \delta v)}$$

The limit of the numerator is $v\dfrac{du}{dx} - u\dfrac{dv}{dx}$

and the limit of the denominator is v^2, since $\delta v \to 0$

Therefore $\qquad \dfrac{dy}{dx} = \dfrac{v\dfrac{\delta u}{\delta x} - u\dfrac{\delta v}{\delta x}}{v^2}$

Example 6.3.1

Differentiate $y = \dfrac{3x}{x-1}$

Using the formula $\dfrac{dy}{dx} = \dfrac{v\dfrac{du}{dx} - u\dfrac{dv}{dx}}{v^2}$ with $u = 3x$ and $v = x - 1$

$$\dfrac{dy}{dx} = \dfrac{\{(x-1) \times 3\} - \{3x \times 1\}}{(x-1)^2}$$

$$= \dfrac{3x - 3 - 3x}{(x-1)^2}$$

$$= \dfrac{-3}{(x-1)^2}$$

Example 6.3.2

Differentiate $y = \dfrac{x^3 + 1}{x^3 - 1}$

Using the same formula with $u = x^3 + 1$ and $v = x^3 - 1$

$$\dfrac{dy}{dx} = \dfrac{\left\{(x^3 - 1) \times 3x^2\right\} - \left\{(x^3 + 1) \times 3x^2\right\}}{(x^3 - 1)^2}$$

$$= \dfrac{3x^5 - 3x^2 - 3x^5 - 3x^2}{(x^3 - 1)^2}$$

$$= \dfrac{-6x^2}{(x^3 - 1)^2}$$

Example 6.3.3

Differentiate $\dfrac{x(x+1)}{x^2-3x+2}$

Let $y = \dfrac{x(x+1)}{x^2-3x+2}$ and use the quotient formula.

Then $\dfrac{dy}{dx} = \dfrac{\left\{(x^2-3x+2)\times(2x+1)\right\}-\left\{(x^2+x)\times(2x-3)\right\}}{(x^2-3x+2)^2}$

$\qquad\quad = \dfrac{(2x^3-5x^2+x+2)-(2x^3-x^2-3x)}{(x^2-3x+2)^2}$

$\qquad\quad = \dfrac{-4x^2+4x+2}{(x^2-3x+2)^2}$

Exercise 6.3

Differentiate the following functions with respect to x.

1 $\dfrac{3}{2x-1}$

2 $\dfrac{1}{1-3x^2}$

3 $\dfrac{x}{x+2}$

4 $\dfrac{x+1}{x+2}$

5 $\dfrac{3x-1}{2x+3}$

6 $\dfrac{x+b}{x-b}$

7 $\dfrac{x-b}{x+b}$

8 $\dfrac{x^2}{x-4}$

9 $\dfrac{x^2}{x^2-4}$

10 $\dfrac{\sqrt{x}}{x+1}$

11 $\dfrac{x+1}{\sqrt{x}}$

12 $\dfrac{\sqrt{x}+1}{\sqrt{x}-1}$

13 $\dfrac{x^3-1}{x^3+1}$

14 $\dfrac{x^2+x-1}{x^2-x+1}$

15 $\dfrac{2x^2-x+1}{3x^2+x-1}$

16 $\dfrac{1+x+x^2}{x}$

17 $\dfrac{2x^4}{a^2 - x^2}$

18 $\dfrac{2x - 3}{2 - 3x}$

19 $\dfrac{x(x - 1)}{x - 2}$

20 $\dfrac{x^{\frac{1}{2}} + 2}{x^{\frac{3}{2}}}$

6.4 Function of a function

To understand the meaning of 'function of a function' consider the trigonometric function $\sin^2 x$, that is $(\sin x)^2$. This function, being the square of $\sin x$, is a function of $\sin x$, just as x^2 is a function of x, or u^2 is a function of u. But $\sin x$ is itself a function of x. Therefore $\sin^2 x$ is a function of $\sin x$, which is a function of x, that is, $\sin^2 x$ is a function of a function of x. Similarly $\sqrt{x^2 + 4x}$ is a function of $x^2 + 4x$, just as $\sqrt{x}$ is a function of x. Therefore $\sqrt{x^2 + 4x}$ is a function of $x^2 + 4x$, which is a function of x.

You can extend the idea of a 'function of a function' further. For example, $\sin^2(\sqrt{x})$ is a function of $\sin \sqrt{x}$, which is a function of $\sqrt{x}$ which is in turn a function of x. In this section, a rule is derived for differentiating a function of a function. However, the example $\sin^2 x$ cannot be taken further at this stage as the rules for differentiating trigonometric functions are in Chapter 8. The algebraic function, $y = (x^2 - 5)^4$ is used as an example for differentiating a function of a function. $(x^2 - 5)^4$ is a function, the fourth power of $x^2 - 5$

This is itself a function of x, if you write $u = x^2 - 5$ then $y = u^4$

Differentiating y with respect to u, you have $\dfrac{dy}{du} = 4u^3$

But you need $\frac{dy}{dx}$, not $\frac{dy}{du}$ Here is a method you can use to find it:

Let δx be an increase in x, leading to corresponding increases δu and δy in u and y. Using the ordinary rules of fractions

$$\frac{\delta y}{\delta x} = \frac{\delta y}{\delta u} \times \frac{\delta u}{\delta x}$$

Now let $\delta x \to 0$. In consequence δu and δy will also approach zero. Then the ratios $\frac{\delta y}{\delta x}, \frac{\delta y}{\delta u}$ and $\frac{\delta u}{\delta x}$ approach the limits $\frac{dy}{dx}, \frac{dy}{du}$ and $\frac{du}{dx}$ respectively. Therefore, using Theorem 3 on limits in Section 2.6:

$$\lim_{\delta x \to 0} \frac{\delta y}{\delta x} = \lim_{\delta x \to 0} \left(\frac{\delta y}{\delta u} \times \frac{\delta u}{\delta x} \right)$$

$$= \lim_{\delta x \to 0} \left(\frac{\delta y}{\delta u} \right) \times \lim_{\delta x \to 0} \left(\frac{\delta u}{\delta x} \right)$$

Therefore $\qquad \dfrac{dy}{dx} = \dfrac{dy}{du} \times \dfrac{du}{dx}$

This result is called the **function of a function rule,** or sometimes simply **the function rule.** Applying the function of a function rule to the example $y = (x^2 - 5)^4$ you find:

$$\frac{dy}{du} = 4u^3$$

and as $u = x^2 - 5 \qquad \dfrac{du}{dx} = 2x$

Using the function of a function rule, $\dfrac{dy}{dx} = \dfrac{dy}{du} \times \dfrac{du}{dx}$

gives $\qquad \dfrac{dy}{dx} = 4u^3 \times 2x$

Therefore $\dfrac{dy}{dx} = 4(x^2 - 5)^3 \times 2x = 8x(x^2 - 5)^3$

Example 6.4.1

Differentiate $y = \sqrt{1 - x^2}$

First write $\sqrt{1 - x^2}$ in index form to obtain $y = (1 - x^2)^{\frac{1}{2}}$

Let $u = 1 - x^2$, so $y = u^{\frac{1}{2}}$

Therefore $\dfrac{dy}{du} = \dfrac{1}{2}u^{-\frac{1}{2}} = \dfrac{1}{2}(1-x^2)^{-\frac{1}{2}}$ and $\dfrac{du}{dx} = -2x$

Since
$$\dfrac{dy}{dx} = \dfrac{dy}{du} \times \dfrac{du}{dx}$$

$$\dfrac{dy}{dx} = \dfrac{1}{2}(1-x^2)^{-\frac{1}{2}} \times (-2x)$$

$$= -x(1-x^2)^{-\frac{1}{2}}$$

$$= \dfrac{-x}{\sqrt{1-x^2}}$$

Nugget – A rule for differentiating a function of a function

In many cases, it is possible to identify a function of the function in terms of an *outer* and an *inner*. For example, if you look at the function $(x^2 - 5)^4$, it has an outer of (something)^4, and has an inner of $x^2 - 5$.

The rule in my head for differentiating this sort of function is:

▶ Differentiate the outer, times the derivative of the inner.

In this example:

▶ differentiating the outer, (something)^4, gives $4 \times$ (something)^3

▶ the derivative of the inner, the derivative of $x^2 - 5$, gives $2x$

Applying this rule gives:

$$y = (x^2 - 5)^4$$

$$\dfrac{dy}{dx} = 4(x^2 - 5)^3 \times 2x$$

As before, read this rule through several times and then, when you think you understand it, see if you can apply it in the worked examples below.

Example 6.4.2

Differentiate $y = (x^2 - 3x + 5)^3$

Let $u = x^2 - 3x + 5$, so $y = u^3$

Therefore $\dfrac{dy}{du} = 3u^2 = 3(x^2 - 3x + 5)^2$ and $\dfrac{du}{dx} = 2x - 3$

Using the function of a function rule:

$$\frac{dy}{dx} = 3(x^2 - 3x + 5)^2 \times (2x - 3)$$
$$= 3(2x - 3)(x^2 - 3x + 5)^2$$

After some practice, you will probably find that you can dispense with the use of 'u' and can write down the result. It is worth practising this shortcut procedure.

Example 6.4.3

Differentiate $y = (3x^2 - 5x + 4)^{\frac{3}{2}}$

Working this example without introducing u, you find

$$\frac{dy}{dx} = \frac{3}{2}(3x^2 - 5x + 4)^{\frac{3}{2}-1} \times \text{derivative of } (3x^2 - 5x + 4)$$
$$= \frac{3}{2}(3x^2 - 5x + 4)^{\frac{1}{2}}(6x - 5)$$

Example 6.4.4

Differentiate $y = (x^2 + 5)\sqrt[3]{x^2 + 1}$

First write this in the form $y = (x^2 + 5)(x^2 + 1)^{\frac{1}{3}}$

Using the product rule,

$$\frac{dy}{dx} = (x^2 + 5) \times \frac{d}{dx}(x^2 + 1)^{\frac{1}{3}} + (x^2 + 1)^{\frac{1}{3}} \times \frac{d}{dx}(x^2 + 5)$$

To find $\dfrac{d}{dx}(x^2 + 1)^{\frac{1}{3}}$ use the function of a function rule. It is better to do this separately, and then substitute its value afterwards.

$$\frac{d}{dx}(x^2 + 1)^{\frac{1}{3}} = \frac{1}{3}(x^2 + 1)^{\frac{1}{3}-1} \times \text{derivative of } (x^2 + 1)$$

$$= \frac{2x}{3}(x^2 + 1)^{-\frac{2}{3}}$$

$$= \frac{2x}{3(x^2 + 1)^{\frac{2}{3}}}$$

Substituting in the expression for $\dfrac{dy}{dx}$,

$$\frac{dy}{dx} = (x^2 + 5) \times \frac{2x}{3(x^2 + 1)^{\frac{2}{3}}} + (x^2 + 1)^{\frac{1}{3}} \times 2x$$

$$= \frac{2x(x^2 + 5)}{3(x^2 + 1)^{\frac{2}{3}}} + 2x(x^2 + 1)^{\frac{1}{3}}$$

If you wish you could simplify this further to get

$$\frac{dy}{dx} = \frac{2x(x^2 + 5) + 6x(x^2 + 1)}{3(x^2 + 1)^{\frac{2}{3}}} = \frac{8x^3 + 16x}{3(x^2 + 1)^{\frac{2}{3}}}$$

Example 6.4.5

Differentiate $y = \dfrac{\sqrt{1 + 3x}}{4x}$

First write this in the form $y = \dfrac{(1 + 3x)^{\frac{1}{2}}}{4x}$

Using the quotient rule,

$$\frac{dy}{dx} = \frac{4x \times \{\text{derivative of } (1 + 3x)^{\frac{1}{2}}\} - (1 + 3x)^{\frac{1}{2}} \times \{\text{derivative of } 4x\}}{(4x)^2}$$

Using the function of a function rule on $\dfrac{d}{dx}(1 + 3x)^{\frac{1}{2}}$

$$\frac{d}{dx}(1 + 3x)^{\frac{1}{2}} = \frac{1}{2}(1 + 3x)^{\frac{1}{2}-1} \times 3 = \frac{3}{2}(1 + 3x)^{-\frac{1}{2}} = \frac{3}{2(1 + 3x)^{\frac{1}{2}}}$$

Substituting in the expression for $\dfrac{dy}{dx}$

$$\frac{dy}{dx} = \frac{4x \times \dfrac{3}{2(1+3x)^{\frac{1}{2}}} - (1+3x)^{\frac{1}{2}} \times 4}{(4x)^2}$$

$$= \frac{\dfrac{6x}{(1+3x)^{\frac{1}{2}}} - 4(1+3x)^{\frac{1}{2}}}{16x^2}$$

$$= \frac{6x - 4(1+3x)}{16x^2(1+3x)^{\frac{1}{2}}}$$

$$= \frac{6x - 4 - 12x}{16x^2(1+3x)^{\frac{1}{2}}}$$

$$= \frac{-6x - 4}{16x^2(1+3x)^{\frac{1}{2}}}$$

$$= \frac{-3x - 2}{8x^2(1+3x)^{\frac{1}{2}}}$$

Exercise 6.4

Differentiate the following with respect to x.

1 $(2x + 5)^2$ **2** $(1 - 5x)^4$ **3** $(3x + 7)^{\frac{1}{3}}$

4 $\dfrac{1}{1 - 2x}$ **5** $(1 - 2x)^2$ **6** $\sqrt{1 - 2x}$

7 $(x^2 - 4)^5$ **8** $(1 - x^2)^{\frac{3}{2}}$ **9** $\sqrt{3x^2 - 7}$

10 $\dfrac{1}{1 - 2x^2}$ **11** $\sqrt{1 - 2x^2}$ **12** $x\sqrt{1 - x^2}$

13 $\dfrac{1}{4 - x}$ **14** $\dfrac{1}{\sqrt{4 - x}}$ **15** $\dfrac{1}{(4 - x)^2}$

16 $\dfrac{1}{x^2 - 1}$ **17** $\dfrac{1}{\sqrt{x^2 - 1}}$ **18** $\dfrac{x}{\sqrt{1 + x^2}}$

19 $\dfrac{x}{\sqrt{1 - x^2}}$ **20** $\sqrt{\dfrac{x}{1 - x}}$ **21** $\sqrt{\left(\dfrac{1 - x}{1 + x}\right)}$

22 $x\sqrt{\left(\dfrac{1-x}{1+x}\right)}$ **23** $\sqrt[3]{x^2+1}$ **24** $\sqrt{a^2+x^2}$

25 $\dfrac{1}{\sqrt{a^2+x^2}}$ **26** $\sqrt{1-x+x^2}$ **27** $(1-2x^2)^n$

28 $\dfrac{x^2}{\sqrt{a^2-x^2}}$ **29** $\left(x+\dfrac{1}{x}\right)^2$ **30** $\dfrac{1}{\sqrt{1+x^3}}$

31 $\dfrac{\sqrt{1+2x}}{x}$ **32** $\dfrac{x}{\sqrt{1+x^2}}$ **33** $\dfrac{\sqrt{1+x^2}}{x}$

34 $\dfrac{1}{\sqrt{2x^2-3x+4}}$ **35** $x^2\sqrt{1-x}$ **36** $\dfrac{\sqrt{1-x^2}}{1-x}$

37 $x\sqrt{2x+3}$

6.5 Differentiating implicit functions

It was stated in Section 1.9 that it frequently happens that, when y is a function of x, the relation between x and y is not explicitly stated, but the two variables occur in an equation from which sometimes y can be determined in terms of x, and sometimes it cannot. Even if you can find y in terms of x it may be in a form in which differentiation may be tedious or difficult.

In these cases, the best method is to differentiate term by term throughout the equation, remembering that in differentiating functions of y, you are differentiating a function of a function.

Example 6.5.1

Find $\frac{dy}{dx}$ from the equation $x^2-y^2+3x=5y$.

Differentiating term by term you find,

$$2x-2y\frac{dy}{dx}+3=5\frac{dy}{dx}$$

At this stage $\frac{dy}{dx}$ has not been found explicitly, but you can find it by solving the above equation for $\frac{dy}{dx}$. Thus collecting terms:

$$2y\frac{dy}{dx} + 5\frac{dy}{dx} = 2x + 3$$

Then $\quad (2y + 5)\frac{dy}{dx} = 2x + 3$

giving $\quad\quad \frac{dy}{dx} = \frac{2x + 3}{2y + 5}$

You can see that the solution gives $\frac{dy}{dx}$ in terms of both x and y. If you know the corresponding values of x and y you can calculate the numerical value of $\frac{dy}{dx}$. Example 6.5.2 shows this.

Example 6.5.2

Find the slope of the tangent to the curve $x^2 + xy + y^2 = 4$ at the point $(2, -2)$.

Differentiating term by term as in Example 6.5.1, and remembering that xy is a product,

$$2x + \left(y + x\frac{dy}{dx}\right) + 2y\frac{dy}{dx} = 0$$

Therefore $\quad (x + 2y)\frac{dy}{dx} = -2x - y$

so $\quad\quad \frac{dy}{dx} = \frac{-2x - y}{x + 2y}$

Therefore, at the point $(2, -2)$

$$\frac{dy}{dx} = \frac{-4 - (-2)}{2 - 4} = 1$$

Therefore the gradient of the tangent at this point is 1 and the slope of the tangent is $45°$.

In questions 1 to 4 find $\frac{dy}{dx}$ from the given implicit functions.

1 $3x^2 + 7xy + 9y^2 = 6$　　　　**2** $(x^2 + y^2)^2 - (x^2 - y^2) = 0$

3 $x^3 + y^3 = 3xy$　　　　　　　**4** $x^n + y^n = a^n$

5 Find the gradient of the tangent to the curve with equation $x^2 + y^2 - 3x + 4y - 3 = 0$ at the point $(1, 1)$.

6.6 Successive differentiation

In Section 6.4, it was stated that the derivative of a function of x is itself a function of x. For example, if

$$y = 3x^4$$

then　　　　　　　　$$\frac{dy}{dx} = 12x^3$$

So $12x^3$ is a function of x, so you can differentiate it with respect to x

$$\frac{d}{dx}(12x^3) = 36x^2$$

This expression is called the **second derivative** or the **second differential coefficient** of the original function. The operation is denoted by

$$\frac{d}{dx}\left(\frac{dy}{dx}\right)$$

The symbol $\frac{d^2y}{dx^2}$ is used to denote the second derivative.

In this symbol the figure '2' in the numerator and denominator is not an index, but shows that y, the original function, has been differentiated twice with respect to x.

Thus, $\frac{d^2y}{dx^2}$ measures the rate at which $\frac{dy}{dx}$ changes with respect to x, just as $\frac{dy}{dx}$ measures the rate at which y is changing with respect to x.

The second derivative is also a function of x. Therefore $\frac{d^2y}{dx^2}$ can also be differentiated with respect to x. The result is the **third derivative** of y with respect to x, and denoted by $\frac{d^3y}{dx^3}$.

6. Some rules for differentiation （83）

Thus you can have a succession of derivatives. You can continue the process of successive differentiation indefinitely, or until one of the derivatives is zero. This is illustrated by the example of $y = x^n$.

Example 6.6.1

Find the successive derivatives of $y = x^n$

$$y = x^n$$

$$\frac{dy}{dx} = nx^{n-1}$$

$$\frac{d^2y}{dx^2} = n(n-1)x^{n-2}$$

$$\frac{d^3y}{dx^3} = n(n-1)(n-2)x^{n-3}$$

If n is a positive integer, you can continue this process until ultimately $n - n$ is reached as the index of x. The derivative becomes $n(n-1)(n-2) \ldots 3.2.1$, that is, factorial n, or $n!$. The next and subsequent coefficients are all zero.

If n is not a positive integer you can continue the process indefinitely.

Example 6.6.2

Find the successive derivatives of $y = x^3 - 7x^2 + 6x + 3$

$$\frac{dy}{dx} = 3x^2 - 14x + 6$$

$$\frac{d^2y}{dx^2} = 6x - 14$$

$$\frac{d^3y}{dx^3} = 6$$

$$\frac{d^4y}{dx^4} = 0$$

6.7 Alternative notation for derivatives

The successive derivatives may be denoted by the following symbols.

▶ When function notation is used
if $f(x)$ or $\phi(x)$ denotes a function of x

$f'(x)$ or $\phi'(x)$ denotes the first derivative

$f''(x)$ or $\phi''(x)$ denotes the second derivative

$f'''(x)$ or $\phi'''(x)$ denotes the third derivative, and so on.

▶ When y notation is used
if y denotes the function of x

y_1 denotes the first derivative

y_2 denotes the second derivative

y_3 denotes the third derivative, and so on.

Sometimes the notation y', y'', y''', ... is used.

6.8 Graphs of derivatives

Successive differentiation of a function of x produces a set of derived functions which are also functions of x. You can draw the graphs of functions and their derived functions. Various relations exist between them.

Example 6.8.1

Consider the function $y = x^2 - 4x + 3$

$$y_1 = 2x - 4$$

$$\text{and } y_2 = 2$$

Figure 6.1 shows the graphs of $y = x^2 - 4x + 3$, $y_1 = 2x - 4$ and $y_2 = 2$. You can see the following relations between the original graph and the graphs of its derivatives.

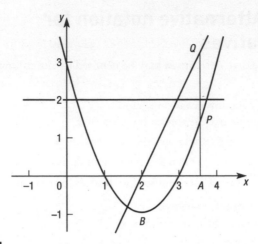

Figure 6.1

▶ Since y_1 – the first derivative – gives the rate of increase of y with respect to x, its value for any given value of x is the gradient at the corresponding point on the graph.

Take a point A on the x-axis where $x = 3.6$. Drawing the ordinate at A, P is the corresponding point on the graph, and Q is the point on the graph of $y_1 = 2x - 4$, the first derivative.

Then the value of the ordinate QA is equal to the gradient of the curve at P. This value is 3.2 units. By calculation, substituting $x = 3.6$ in $y_1 = 2x - 4$, the derivative, gives $(2 \times 3.6) - 4 = 3.2$.

▶ The graph of the second derivative $y_2 = 2$, being parallel to the x-axis and having a constant value, shows that the gradient of $y_1 = 2x - 4$ is constant, namely 2.

▶ At the lowest point B on the curve $y = x_2 - 4x + 3$ the y-value at the corresponding point on the first derivative $y_1 = 2x - 4$, is zero; it cuts the x-axis at this point. Thus the gradient of the original function is zero when $x = 2$. A tangent drawn to the graph at B will be parallel to the x-axis.

▶ For values of x less than 2, the function $x_2 - 4x + 3$ is decreasing while its derivative is negative. For values greater than 2, $x_2 - 4x + 3$ is increasing and the derivative is positive.

Exercise 6.6

In questions 1 to 8, write down the first, second and third derivatives of the given function of x.

1 $x^2(x-1)$ **2** x^{2b}

3 $5x^4 - 3x^3 + 2x^2 - x + 1$ **4** $10x^5 - 4x^3 + 5x - 2$

5 $\dfrac{1}{x}$ **6** $\sqrt{x}$

7 $\sqrt{2x+1}$ **8** $\dfrac{1}{x^2}$

9 Use the relation $\dfrac{1}{a^2 - x^2} = \dfrac{1}{2a}\left(\dfrac{1}{a+x} + \dfrac{1}{a-x}\right)$ to find the nth derivative of $\dfrac{1}{a^2 - x^2}$

10 Let $f(x) = 6x^2 - 5x + 3$. Find $f'(0)$. For what value of x is $f'(x) = 0$? To what point of the graph of $f(x)$ does this correspond?

11 Let $f(x) = x^3 - 5x^2 + 7$. Find $f'(1)$ and $f''(2)$. For what values of x does $f'(x)$ vanish?

12 Find the values of x for which $f(x) = \dfrac{1}{3}x^3 - \dfrac{5}{2}x^2 + 6x + 1$ has zero gradient. For what value of x is the gradient of $f'(x)$ equal to zero? To what values of $f'(x)$ does this value of x correspond?

Key ideas

▶ The procedure for differentiating a *sum* or a *difference* is straightforward – simply differentiate each term separately.

▶ The first strategy for *differentiating a product* is to try to multiply it out and then use the rule for differentiating a sum. Failing that, use the special rule for differentiating a product, which is:

if $y = u \times v$, $\dfrac{dy}{dx} = u\dfrac{dv}{dx} + v\dfrac{du}{dx}$

- ▶ The special rule for *differentiating a quotient* is:

  if $y = \dfrac{u}{v}$, $\dfrac{dy}{dx} = \dfrac{\left(v \dfrac{du}{dx} - u \dfrac{dv}{dx} \right)}{v^2}$

- ▶ A *function of a function* is something like $\sin^2 x$ (i.e. $\sin x$)^2. Since $\sin x$ is a function of x, so its square, $\sin^2 x$, is a function of a function of x. There is a special rule for finding the derivative of a function of a function (see Section 6.4).

- ▶ There is a method for finding the *derivative of an implicit function* (see Section 6.5).

- ▶ The derivative $\frac{dy}{dx}$ is called the *first derivative* of y with respect to x. If you differentiate $\frac{dy}{dx}$ with respect to x, you get the *second derivative*, written $\frac{d^2y}{dx^2}$. Differentiating a second derivative gets a *third derivative* $\frac{d^3y}{dx^3}$ and so on.

7

Maxima, minima and points of inflexion

In this chapter you will learn:

▶ *how to find points on graphs where the y-values are greatest or least*

▶ *how to distinguish between such points without using a diagram*

▶ *how to use this knowledge in applications.*

7.1 Sign of the derivative

The sign of the derivative was introduced in Section 5.5 and in Example 6.8.1. In this chapter it is examined in more detail.

Let y be a continuous function of x, and let $\frac{dy}{dx}$ be positive for a particular value of x, say $x = a$. Then, at $x = a$, since $\tan \theta = \frac{dy}{dx}$ $\theta > 0$ and the graph of y against x appears as in Figure 7.1.

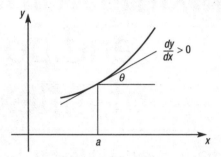

Figure 7.1

Therefore, if $\frac{dy}{dx}$ is positive, then y increases as x increases.

Similarly, if $\frac{dy}{dx}$ is negative, then y decreases as x increases.

Nugget – The differential coefficient and gradient

In this chapter you will be looking at the link between the gradient of the general function $y = f(x)$ at various points on it and the corresponding values of the differential coefficient of the function, $\frac{dy}{dx}$

The two main points to be clear about at this stage are:

▶ the differential coefficient $\frac{dy}{dx}$ *is* the gradient

▶ it therefore follows that at those points on the curve where the gradient is positive (i.e. where it slopes from bottom left to top right), the value of $\frac{dy}{dx}$ is positive; also, where the gradient is negative (i.e. where it slopes from top left to bottom right), the value of $\frac{dy}{dx}$ is negative.

Figure 7.2 shows functions which are increasing in various ways.

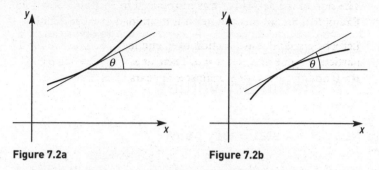

Figure 7.2a   **Figure 7.2b**

In Figure 7.2a the shape of the curve is **concave upwards** and the function is **increasing**. Examples of curves with this property are $y = x^2$, $y = 10^x$ and $y = \tan x$ for $0 < x < \frac{1}{2}\pi$

In Figure 7.2b the shape of the curve is **concave downwards** and the function is **increasing**. Examples of curves with this property are $y = \sqrt{x}$, $y = \ln x$ and $y = \sin x$ for $0 < x < \frac{1}{2}\pi$

In both cases, the tangent to the curve makes an acute angle θ with the positive x-axis, and $\tan \theta$ is positive.

Figure 7.3 shows functions which are decreasing in various ways.

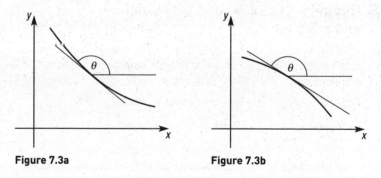

Figure 7.3a   **Figure 7.3b**

In Figure 7.3a the shape of the curve is **concave upwards** and the function is **decreasing**. Examples of curves with this property are $y = x^2$ for $x < 0$ and $y = \frac{1}{x}$

In Figure 7.3b the curve is **concave downwards** and the function is **decreasing**. Examples of curves with this property are $y = \sin x$ for $\frac{1}{2}\pi < x < \pi$ and $y = -x^2$ for positive values of x

In both cases, the tangent to the curve makes an obtuse angle θ with the positive x-axis, and $\tan \theta$ is negative.

7.2 Stationary values

Nugget – A stationary point

A *stationary point* on a graph is where the rate of change is zero. What this means is that the value of $\frac{dy}{dx}$ equals zero at this point. It is called a stationary point because this is where the function momentarily *stops* increasing or decreasing.

Usually, but not always, this stationary point will be either a maximum point (on the graph this will look like the peak of a hill) or a minimum point (on the graph this will look like the bottom of a valley).

Two of the cases illustrated in Section 7.1, namely Figures 7.1 and 7.2a occur in the graph of $y = x^2 - 4x + 3$, shown in Figure 6.1, and repeated as Figure 7.4.

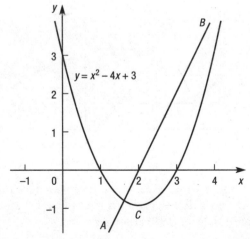

Figure 7.4

Since
$$y = x^2 - 4x + 3$$
$$\frac{dy}{dx} = 2x - 4$$

The graph of $\frac{dy}{dx} = 2x - 4$ is represented in Figure 7.4 by the line AB.

You can see the following facts from Figure 7.4.

▶ As x increases from $-\infty$ to 2, the value of $\frac{dy}{dx}$, represented by AB, is negative, so y is decreasing.

▶ As x increases from 2 to ∞, the value of $\frac{dy}{dx}$, represented by AB, is positive, so y is increasing.

▶ At C, the curve stops decreasing and starts to increase. Thus, when $x = 2$, the value of y is momentarily not changing, but is stationary. The gradient there is zero, and the rate of change is also zero. The straight line AB cuts the x-axis at the point $x = 2$.

At $x = 2$, the function $y = x^2 - 4x + 3$ is said to have a **stationary value**, and C is called a **stationary point** on the curve.

In Figure 7.5, the graph of $y = 3 + 2x - x^2$, and its derivative $\frac{dy}{dx} = 2 - 2x$, which is the straight line AB, have been drawn.

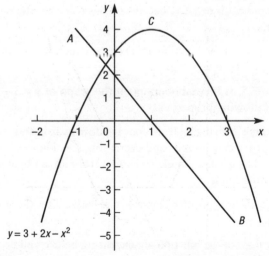

Figure 7.5

You can see from Figure 7.5 that:

▶ when $x < 1$ and $\frac{dy}{dx}$ is positive, y is increasing

▶ when $x > 1$ and $\frac{dy}{dx}$ is negative, y is decreasing

▶ at $x = 1$, at C, y has stopped increasing and started to decrease. The value of the function at C is stationary, and the curve has a stationary point.

Definition: A **stationary point** is a point on a graph for which $\frac{dy}{dx} = 0$.

7.3 Turning points

Comparing the stationary points of Figures 7.4 and 7.5, the following differences appear.

In Figure 7.4, at the stationary point the graph of $y = x^2 - 4x + 3$ has the following properties.

▶ The curve is changing from concave upwards and decreasing to concave upwards and increasing (see Figures 7.2a and 7.3a). The gradient angle θ changes from an obtuse angle to an acute angle.

▶ The gradient is negative before and positive after the stationary point.

▶ The values of the function are decreasing before and increasing after the stationary point.

In Figure 7.5, at the stationary point the graph of $y = 3 + 2x - x^2$ has the following properties.

▶ The curve is changing from concave downwards and increasing to concave downwards and decreasing (see Figures 7.2b and 7.3b). The gradient angle θ changes from an acute angle to an obtuse angle.

▶ The gradient is positive before and negative after the stationary point.

▶ The values of the function are increasing before and decreasing after the stationary point.

At both these stationary points:

▶ the function decreases before and increases after, or *vice versa*

▶ $\frac{dy}{dx} = 0$, and changes sign.

Definition: Points on curves which have these two properties are called **turning points**. You will see later in Section 7.8 that not all stationary points are turning points.

Both stationary points and turning points have the property that $\frac{dy}{dx} = 0$.

Example 7.3.1

For what value of x is there a turning point on the graph of $y = 2x^2 - 6x + 9$?

$$y = 2x^2 - 6x + 9$$

$$\frac{dy}{dx} = 4x - 6$$

For a stationary point $\frac{dy}{dx} = 0$, so $4x - 6 = 0$, or $x = 1\frac{1}{2}$.

For $x < 1\frac{1}{2}$, $\frac{dy}{dx}$ is negative, so the function is decreasing.

For $x > 1\frac{1}{2}$, $\frac{dy}{dx}$ is positive, so the function is increasing.

As the function is decreasing before the stationary point and increasing after it, there is a turning point at $x = 1\frac{1}{2}$.

You are recommended to draw the graph on your calculator.

Example 7.3.2

Find the turning points of $y = 1 - 2x - x^2$

$$y = 1 - 2x - x^2$$

$$\frac{dy}{dx} = -2 - 2x$$

For a stationary point $\frac{dy}{dx} = 0$, so $-2 - 2x = 0$, or $x = -1$.

For $x < -1$, $\frac{dy}{dx}$ is positive, so the function is increasing.

7.4 Maximum and minimum values

There is an important and obvious difference between the turning points of the curves in Section 7.2, that is $y = x^2 - 4x + 3$ and $y = 3 + 2x - x^2$ in Figures 7.4 and 7.5.

In $y = x^2 - 4x + 3$, the turning point C is the lowest point on the curve, that is, the point at which y has its least value. If points on the curve are taken close to and on either side of C, the value of the function at each of them is greater than at C, the turning point.

Such a point is called a **local minimum**, or often just a **minimum**, and the function is said to have a minimum value for the corresponding value of x.

Notice that the values of the function decrease before the minimum point, and increase after it.

In $y = 3 + 2x - x^2$ the turning point C is the highest point on the curve, that is, the point at which y has its greatest value. If points on the curve are taken close to and on either side of C, the value of the function at each of them is less than at C, the turning point.

Such a point is called a **local maximum**, or often just a **maximum**, and the function is said to have a maximum value for the corresponding value of x.

Notice that the values of the function increase before the maximum point, and decrease after it.

The values of the function at the local maximum and local minimum points, while greater than or less than the values at points close to them on the curve, are not necessarily the greatest and least values respectively which the function may have.

Example 7.4.1 shows a case in which the same graph contains both local maxima and local minima.

Example 7.4.1

Find the turning points of the function $y = (x - 1)(x - 2)(x - 3)$ and sketch the curve.

Here are some pointers about the general features of the function.

The function is zero when $x - 1 = 0$, when $x - 2 = 0$ and when $x - 3 = 0$, that is, when $x = 1$, $x = 2$ and $x = 3$.

The curve therefore cuts the x-axis for these values of x. If the function is continuous, then between two consecutive values for which the curve cuts the x-axis there must be a turning point. Therefore:

▶ there is a turning point between $x = 1$ and $x = 2$

▶ there is a turning point between $x = 2$ and $x = 3$.

Notice also that:

▶ when $x < 1$, y is negative

▶ when $1 < x < 2$, y is positive

▶ when $2 < x < 3$, y is negative

▶ when $x > 3$, y is positive.

Putting these results together, there is a local maximum between $x = 1$ and $x = 2$, and a local minimum between $x = 2$ and $x = 3$.

Using a graphics calculator or a spreadsheet, you can verify these features by drawing the graph, shown in Figure 7.6.

Working algebraically, you find after expanding the brackets that:

$$y = x^3 - 6x^2 + 11x - 6$$
$$\frac{dy}{dx} = 3x^2 - 12x + 11$$

Turning points occur when $\frac{dy}{dx} = 0$, so $3x^2 - 12x + 11 = 0$

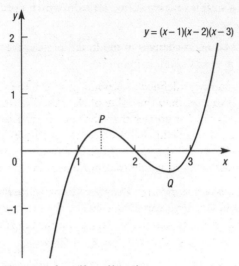

Figure 7.6 Graph of $y = (x-1)(x-2)(x-3)$

Solving this equation you find that $x = 1.42$ and $x = 2.58$, both values correct to three significant figures.

For these values of x there are turning points, marked as P and Q on Figure 7.6.

Substituting these values for x to find y, you find that $y = 0.385$, the point P in Figure 7.6, and $y = -0.385$, the point Q in Figure 7.6.

Therefore y has a local maximum of 0.385 when $x = 1.42$, and a local minimum of -0.385 when $x = 2.58$.

7.5 Which are maxima and which are minima?

In Example 7.4.1 it was possible to decide which turning point was a local maximum and which was a local minimum by thinking about the shape of the graph. This method is not always easy, so you need an algebraic method which you can use quickly to determine which are local maxima and which are local minima.

Three methods are available. They all follow from previous conclusions.

Method 1: Looking at changes in the function near the turning point

A local maximum was defined as a point at which the value of the function is greater than the value of the function for values of x slightly less than or greater than that of the turning point. A local minimum was defined as a point at which the value of the function is less than the value of the function for values of x slightly less than or greater than that of the turning point.

Method 1 consists in applying these definitions. Values of x slightly less and slightly greater than that at the turning point are substituted in the function. By comparing the results you can decide which of the above definitions is satisfied.

You can express this more generally in the following way:

Let $f(x)$ be a function of x, and let a be the value of x at a turning point. Let h be a small positive number. Then $f(a + h)$ is the value of the function at a point slightly greater than a, and $f(a - h)$ is the value of the function at a point slightly less than a. For a local maximum: $f(a)$ is greater than both $f(a + h)$ and $f(a - h)$. For a local minimum: $f(a)$ is less than both $f(a + h)$ and $f(a - h)$.

Method 2: Changes in the value of the derivative before and after the turning point

At a local maximum, the function increases before and decreases after the turning point. This will happen if the gradient is positive before the turning point and negative after the turning point.

Method 2 consists in checking the sign of the gradient at values of x just a little less than and a little greater than the value of x at the turning point. If the sign of the gradient is changing from positive to negative as you pass the turning point, it will be a local maximum.

Similarly, if $\frac{dy}{dx}$ is changing from negative to positive, then the turning point will be a local minimum.

Nugget – Positive, nought, negative

It is useful to keep in mind a mental picture of what a maximum and a minimum look like and what the gradient is doing on either side of them. Look at the two pictures and ask yourself what the gradient is doing as you pass through each turning point. As you can see, for a maximum the gradient starts positive, becomes zero at the turning point and then becomes negative. In summary, the behaviour of the gradient around a local maximum point is 'positive, nought, negative'. The reverse is true for a minimum point, whose gradient goes 'negative, nought, positive'.

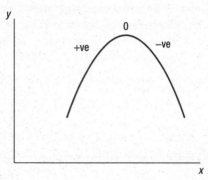

Figure 7.7 The pattern for a maximum: 'positive, nought, negative'

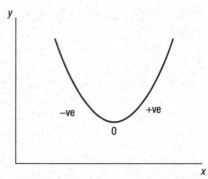

Figure 7.8 The pattern for a minimum: 'negative, nought, positive'

Method 3: The sign of the second derivative

Just as the value of $\frac{dy}{dx}$ tells you whether y is increasing or decreasing, so the value of $\frac{d^2y}{dx^2}$ tells you whether $\frac{dy}{dx}$ is increasing or decreasing. If $\frac{d^2y}{dx^2}$ is positive, then by applying the results of Section 7.1, $\frac{dy}{dx}$ increases as x increases.

Suppose that $\frac{d^2y}{dx^2}$ is positive at a turning point. Then $\frac{dy}{dx}$ is increasing at the turning point. But $\frac{dy}{dx} = 0$ at a turning point. Therefore, as $\frac{dy}{dx}$ is increasing at the point for which $\frac{dy}{dx} = 0$, it must be changing from negative to positive. Therefore, referring to Method 2, the turning point is a local minimum. Similarly, if $\frac{d^2y}{dx^2}$ is negative at a turning point, the turning point is a local maximum.

Beware! If $\frac{d^2y}{dx^2} = 0$ at a stationary point, it could be a local maximum, a local minimum or neither of these, and this method fails. This special case is discussed in Section 7.8.

Nugget – The second derivative

You may remember from the previous chapter that the second derivative $\frac{d^2y}{dx^2}$ was the derivative of the derivative of the function. But what does this second derivative actually mean in terms of the graph? Let's start with the first derivative

The *first* derivative $\frac{dy}{dx}$ gives the rate of change of the function as you travel along its graph from left to right – in other words, its gradient. If the gradient is positive, so too will be the value of the first derivative; if the gradient is negative, so the value of the first derivative will be negative.

The *second* derivative $\frac{d^2y}{dx^2}$ gives the rate of change of the gradient.

If the gradient is increasing (concave upwards), the second derivative will be positive and if the gradient is decreasing (concave downwards), the second derivative will be negative.

7.6 A graphical illustration

The three methods of Section 7.5 can be illustrated by the function of Example 7.4.1, $y = (x - 1)(x - 2)(x - 3)$, or $y = x^3 - 6x^2 + 11x - 6$.

Let $y = f(x)$ so
$$f(x) = x^3 - 6x^2 + 11x - 6$$
$$f'(x) = 3x^2 - 12x + 11$$
$$f''(x) = 6x - 12$$

All three graphs are shown in Figure 7.9.

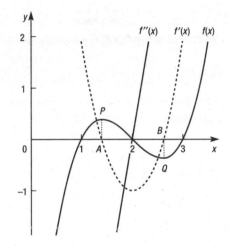

Figure 7.9

METHOD 1

You can see that values of the function near the local maximum P are all less than the value at P. Similarly, the values of the function near the local minimum Q are all greater than the value at Q.

METHOD 2

You can see that the dashed curve, $f'(x)$, has the properties that $f'(x) = 0$ and that $f'(x)$ moves from positive to negative at A, showing that P is a local maximum. Similarly, at B $f'(x) = 0$ and $f'(x)$ moves from negative to positive, showing that Q is a local minimum.

METHOD 3

At A $f''(x)$ is negative, so P is a local maximum. Similarly, at B $f''(x)$ is positive, so P is a local minimum.

In practice, Methods 2 and 3 are easier to use than Method 1. But remember that Method 3 is useless if $f''(x) = 0$ at the turning point.

7.7 Some worked examples

Example 7.7.1

Find any local maxima or minima for the function $y = 2x^2 - 6x + 3$.

$$\frac{dy}{dx} = 4x - 6$$

For a turning point $\frac{dy}{dx} = 0$, so $4x - 6 = 0$, giving $x = 1.5$

Therefore there is a turning point when $x = 1.5$

To determine whether this is a local maximum or a local minimum, we first use Method 2:

Since
$$\frac{dy}{dx} = 4x - 6 = 4(x - 1.5)$$

if $x < 1.5$, then $\frac{dy}{dx}$ is negative

and if $x > 1.5$, $\frac{dy}{dx}$ is positive

Therefore by Method 2, y has a local minimum at $x = 1.5$

Using Method 3:
$$\frac{d^2y}{dx^2} = 4$$

Since this is positive, whatever the value of x, by Method 3, y has a local minimum at $x = 1.5$

Example 7.7.2

Find any local maxima or minima for the function $y = 5 - x - x^2$ and distinguish between them.

$$\frac{dy}{dx} = -1 - 2x$$

For a turning point $\frac{dy}{dx} = 0$, so $-1 - 2x = 0$, giving $x = -0.5$.

Using Method 2, since $\dfrac{dy}{dx} = -2(x + 0.5)$

if $x < -0.5$, then $\frac{dy}{dx}$ is positive

and if $x > -0.5$, then $\frac{dy}{dx}$ is negative

Therefore by Method 2, y has a local maximum at $x = -0.5$

Using Method 3: $\qquad\qquad\qquad \dfrac{d^2y}{dx^2} = -2$

Since this is negative, whatever the value of x, by Method 3, y has a local maximum at $x = -0.5$.

Example 7.7.3

Find the turning points of $y = x^3 - 6x^2 + 9x - 2$ and distinguish between them.

$$y = x^3 - 6x^2 + 9x - 2$$
$$\frac{dy}{dx} = 3x^2 - 12x + 9$$
$$\frac{d^2y}{dx^2} = 6x - 12$$

For a turning point $\frac{dy}{dx} = 0$, so $3x^2 - 12x + 9 = 0$

Therefore $3(x^2 - 4x + 3) = 0$, so $x = 3$ or 1

Therefore there are turning points when $x = 3$ and 1

Using Method 3, when $x = 1$, $\frac{d^2y}{dx^2} = 6 \times 1 - 12 = -6$, so this is a local maximum.

When $x = 3$, $\frac{d^2y}{dx^2} = 6 \times 3 - 12 = 6$, so this is a local minimum.

You can find the corresponding y-values by substituting $x = 1$ and $x = 3$ in the equation $y = x^3 - 6x^2 + 9x - 2$, giving 2 and −2.

Therefore, there is a local maximum at (1, 2) and a local minimum at (3, −2).

Example 7.7.4

When a body is projected vertically upwards with a velocity of 7 ms^{-1} the height reached by the body after t seconds is given by the formula $s = 7t - 4.9t^2$. Find the greatest height to which the body will rise, and the time taken.

Notice that s is a function of t given by $s = 7t - 4.9t^2$

Therefore
$$\frac{ds}{dt} = 7 - 9.8t$$

When s is greatest, $\frac{ds}{dt} = 0$, giving $7 - 9.8t = 0$ and $t = \frac{1}{1.4}$

Since $\frac{d^2s}{dt^2} = -9.8$, which is always negative, this is a maximum value for the height.

Substituting $t = \frac{1}{1.4}$ into $s = 7t - 4.9t^2$ gives $s = 2.5$, so the maximum height is 2.5 m.

Example 7.7.5

The cost, £C per mile, of a cable is given by $C = \frac{120}{x} + 600x$ where $x \text{ cm}^2$ is its cross-sectional area. Find the area of cross-section for which the cost is least, and the least cost per mile.

As
$$C = \frac{120}{x} + 600x = 120x^{-1} + 600x$$

$$\frac{dC}{dx} = 120x^{-2} + 600 = \frac{120}{x^2} + 600$$

For a maximum or minimum cost $\frac{dC}{dx} = 0$, so $-\frac{120}{x^2} + 600 = 0$, giving $x^2 = 0.2$.

Therefore $x = \pm \sqrt{0.2} = \pm 0.447$ (approximately).

The negative root has no meaning in this context, and you should ignore it.

To discover whether $x = \sqrt{0.2}$ corresponds to a maximum or a minimum, use Method 3.

Then
$$\frac{d^2C}{dx^2} = 240x^{-3} = \frac{240}{x^3}$$

so when $x = \sqrt{0.2}$ this is positive, giving a minimum cost.

Thus, the minimum cost is $C = \frac{120}{\sqrt{0.2}} + 600 \times \sqrt{0.2} = £537$ when

the cross-sectional area is 0.447 cm², correct to three significant figures.

Example 7.7.6

A cylindrical gas holder is to be constructed so that its volume is a constant V m³. Find the relation between the radius of its base and its height so that its surface area, not including the base, is the least possible. Find also the radius r m of the base in terms of V.

Let h m be the height of the gas holder, and let A m² be its surface area excluding the base.

Then, using the formulae for the area and volume of a cylinder:

$$A = \pi r^2 + 2\pi r h$$
$$V = \pi r^2 h$$

These equations have two independent variables, r and h. Eliminate one of them, in this case h, between the two equations to obtain an equation for A in terms of r and V. (Remember that V is a constant.)

From the second equation $h = \frac{V}{\pi r^2}$

Substituting this in the first equation:

$$A = \pi r^2 + 2\pi r \times \frac{V}{\pi r^2} = \pi r^2 + \frac{2V}{r}$$

Now A is a function of r so, differentiating

$$\frac{dA}{dr} = 2\pi r - \frac{2V}{r^2}$$

Since A is to be a minimum $\frac{dA}{dr} = 0$, so $2\pi r - \frac{2V}{r^2} = 0$

Solving this equation for r gives $\pi r = \frac{V}{r^2}$ or $V = \pi r^3$

Therefore $r = \sqrt[3]{\dfrac{V}{\pi}}$

Also since $V = \pi r^2 h$, $\pi r^3 = \pi r^2 h$ and $h = r$

Finally, since $\frac{d^2V}{dr^2} = 2\pi + \frac{4V}{r^3}$, it is positive when r is positive, and so this value of r gives a minimum surface area.

Note that you should not differentiate the equation $A = \pi r^2 + 2\pi rh$ as it stands. As this equation for A contains two independent variables r and h you need to eliminate one of them to leave just one independent variable. It may help to keep track of which letters represent variables and which represent constants in such situations.

7.8 Points of inflexion

Imagine that you are a train driver, driving a train along a track which is the graph of a function $y = f(x)$. As you go along the curved track, sometimes you are turning to the left, and sometimes you are turning to the right. Occasionally you change from turning to the left and start turning to the right, and *vice versa*. Such points are called points of inflexion. Figure 7.10 gives examples of points of inflexion. Notice that in some of them the gradient $f'(x)$ is zero, and in others it is not.

Look first at the examples on the left. In all of them $f'(x)$ is decreasing and then starts to increase again. In the top example the least value taken by $f'(x)$ is positive; in the middle example the least value taken by $f'(x)$ is zero; in the bottom example, the least value taken by $f'(x)$ is negative. Thus $f'(x)$ has a minimum value at the point of inflexion.

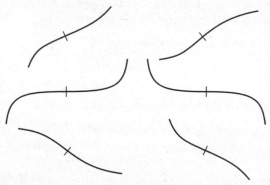

Figure 7.10 Some examples of points of inflexion

A similar situation occurs in the right-hand examples. This time $f'(x)$ is increasing and then starts to decrease. In the top example the greatest value taken by $f'(x)$ is positive; in the middle example the greatest value taken by $f'(x)$ is zero; in the bottom example, the greatest value taken by $f'(x)$ is negative. Thus $f'(x)$ has a maximum value at the point of inflexion.

It is now possible to give a more general definition.

Definition: A **point of inflexion** on a graph of $y = f(x)$ is a point at which the gradient takes either a local maximum value or a local minimum value.

You could also say that a point of inflexion occurs where a curve changes from being concave upwards to concave downwards, or *vice versa*.

In all the cases $f'(x)$ is a stationary point, so $f''(x) = 0$ at a point of inflexion. In the cases where both $f'(x) = 0$ and $f''(x) = 0$, the situation is more complicated, and it is better to go back to first principles to find out what is happening. Thus, when $f'(x) = 0$ and $f''(x) = 0$, use Method 2 in Section 7.5 to determine whether the gradient $f'(x)$ has a maximum or a minimum value.

Notice that the point of inflexion with a horizontal tangent is the special case mentioned at the end of Section 7.5. A stationary point could be a local maximum, a local minimum or a point of inflexion with a horizontal tangent. A turning point is specifically a local maximum or a local minimum.

Example 7.8.1

Return to the curve $y = x^3 - 6x^2 + 11x - 6$ discussed in Example 7.4.1, and shown in Figure 7.9.

You can see that the graph of $f''(x)$ meets the x-axis at $(2, 0)$ and that $f'(2) = -1$. Thus the point $x = 2$ is a point of inflexion.

Example 7.8.2

Find the stationary point of $y = x^3$ and determine its nature.

For $y = x^3$, $\frac{dy}{dx} = 3x^2$ and $\frac{d^2y}{dx^2} = 6x$

For a stationary point $\frac{dy}{dx} = 0$, so $3x^2 = 0$ giving $x = 0$.

When $x = 0$, $\frac{d^2y}{dx^2} = 0$, the situation occurs in which at $x = 0$, both $\frac{dy}{dx} = 0$ and $\frac{d^2y}{dx^2} = 0$.

This is the situation in which it is better to go back to first principles to determine the type of stationary point which occurs at $x = 0$.

When $x < 0$, $\frac{dy}{dx}$ is positive, and when $x > 0$, $\frac{dy}{dx}$ is also positive.

Since the gradient has changed from positive to zero to positive, the gradient has a minimum value, and $x = 0$ is therefore a point of inflexion on the graph of $y = x^3$.

Test yourself

1 Draw the graph of $y = x^2 - 2x$, using equal scales on both axes. Find $\frac{dy}{dx}$ and obtain its value when $x = -1, 0, 1, 2, 3$, checking the values from the graph. For what value of x is there a turning point on the curve? Is this a local maximum or a local minimum? What is the sign of $\frac{d^2y}{dx^2}$?

2 Draw the graph of $y = 3x - x^2$, using equal scales on both axes. Find $\frac{dy}{dx}$ and calculate its value when $x = 0, 1, 2, 3$. For what value of x is $\frac{dy}{dx} = 0$? What is the sign of $\frac{d^2y}{dx^2}$ for the same value of x? Is the function a local maximum or a local minimum for this value of x?

3 Find the turning points for the following functions, and determine whether the function has a local maximum or a local minimum in each case.

a $4x^2 - 2x$ b $x - 1.5x^2$

c $x^2 + 4x + 2$ d $2x^2 + x - 1$

4 Find the local maximum and minimum values of the following functions, and state the corresponding values of x.

    a $x^3 - 12x$ **b** $2x^3 - 9x^2 + 12x$

    c $x^3 - 6x^2 + 12$ **d** $4x^3 + 9x^2 - 12x + 13$

5 Find the local maximum and minimum values of $(x + 1)(x - 2)^2$ and the corresponding values of x.

6 Find the local maximum and minimum values of $4x + \frac{1}{x}$.

7 Divide 10 into two parts such that their product is a maximum.

8 A particle is projected at an angle θ with initial velocity u. Its horizontal displacement x and its vertical displacement y are related by the equation $y = x \tan\theta - \frac{gx^2}{2u^2 \cos^2\theta}$. Find the maximum height to which the particle rises, and find how far it has then travelled horizontally. (The constant g is the acceleration due to gravity.)

9 A closed cylindrical drum is to be manufactured to contain 40 m^3 using the minimum possible amount of material. Find the ratio of the height of the drum to the diameter of its base.

10 An open tank of capacity 8 m^3 is to be made of sheet iron. The tank has a square base, and its sides are perpendicular to its base. The smallest possible amount of metal is to be used. Find the dimensions of the tank.

11 If $\frac{ds}{dt} = 4.8 - 3.2t$ and $s = 5$ when $t = 0.5$, express s as a function of t and find its maximum value.

12 If $H = pV$ and $p = 3 - \frac{1}{2}V$, find the maximum value of H.

13 A rectangular sheet of tin, 30 cm × 24 cm, has four equal squares cut out at the corners, and the sides are then turned up to form a rectangular box. What must be the length of the side of each square cut away, so that the volume of the box is the maximum possible?

14 The strength of a rectangular beam of given length is proportional to bd^3, where b is the breadth and d the depth. The cross section of the beam has a perimeter of 4 m. Find the breadth and depth of the strongest beam.

15 Find the x-values of a maximum value, a minimum value and a point of inflexion for $y = 2x^3 + 3x^2 - 36x + 10$.

16 Find the maximum and minimum values of $y = x(x^2 - 1)$. Find also the gradient at the point of inflexion.

17 Find the point of inflexion on the graph of $y = 3x^3 - 4x + 5$.

18 The displacement s travelled by a body propelled vertically upwards in time t is given approximately by the formula

$$s = 120t - 4.9t^2$$

Find the greatest height which the body will reach, and the time taken to reach it.

19 The bending moment M of a beam, supported at one end, at a distance x from one end is given by the formula

$$M = \tfrac{1}{2}wlx - \tfrac{1}{2}wx^2$$

where l is the length and w is the uniform load per unit length. Find the point on the beam at which the bending moment is a maximum.

Key ideas

▶ If y increases as x increases, $\frac{dy}{dx}$ is positive and the graph of the function has a *positive gradient*.

▶ If y decreases as x increases, $\frac{dy}{dx}$ is negative and the graph of the function has a *negative gradient*.

▶ A *stationary point* occurs where the value of y is momentarily unchanging; there is therefore no rate of change and $\frac{dy}{dx}$ is zero.

▶ A *local maximum* shows on the graph as a peak. It occurs where $\frac{dy}{dx}$ is zero, and the gradient is positive just before and negative just after. Also $\frac{d^2y}{dx^2}$ will be negative at this point, signifying a negative rate of change of gradient.

▶ A *local minimum* shows on the graph as a valley. It occurs where $\frac{dy}{dx}$ is zero, and the gradient is negative just before and positive just after. Also $\frac{d^2y}{dx^2}$ will be positive at this point, signifying a positive rate of change of gradient.

▶ A *point of inflexion* shows on the graph where it changes from concave up to concave down or vice versa. It occurs where the gradient is momentarily unchanging; this is where $\frac{d^2y}{dx^2} = 0$ and its sign is changing.

8

Differentiating the trigonometric functions

In this chapter you will learn:

▶ *how to differentiate the trigonometric functions, sin x, cos x and tan x*

▶ *that it is necessary to measure angles in radians*

▶ *how to differentiate the inverse trigonometric functions, sin⁻¹ x, cos⁻¹ x and tan⁻¹ x.*

8.1 Using radians

When you differentiate trigonometric functions, often called circular functions, you must remember that the angle must be in radians. Thus, when you find the derivative of $\sin \theta$ (that is, the rate of increase of $\sin \theta$ with respect to θ), the angle θ will be in radians unless it is specifically indicated to the contrary.

If you are not confident about the use of radians, you should revise them before continuing with this chapter.

8.2 Differentiating $\sin x$

Let $y = \sin x$

Let δx be an increase in x

Let δy be the corresponding increase in y

Then
$$y + \delta y = \sin(x + \delta x)$$

Thus
$$\delta y = \sin(x + \delta x) - \sin x$$

Dividing by δx
$$\frac{\delta y}{\delta x} = \frac{\sin(x + \delta x) - \sin x}{\delta x}$$

The next step is to find the limit of the right-hand side as $\delta x \to 0$. This requires some trigonometric manipulation.

First change the numerator of the right-hand side from a difference into a product by using the trigonometric formula:

$$\sin P - \sin Q = 2 \cos \frac{P+Q}{2} \sin \frac{P-Q}{2}$$

where $x + \delta x$ takes the place of P and x the place of Q

$$\frac{\delta y}{\delta x} = \frac{2 \cos \dfrac{(x + \delta x) + x}{2} \sin \dfrac{(x + \delta x) - x}{2}}{\delta x}$$

$$= \frac{2 \cos \left(x + \dfrac{\delta x}{2} \right) \sin \dfrac{\delta x}{2}}{\delta x}$$

or, re-arranging

$$\frac{\delta y}{\delta x} = 2\, \cos\left(x + \frac{\delta x}{2}\right) \frac{\sin \dfrac{\delta x}{2}}{\delta x}$$

Transferring the numerical factor from the numerator to the denominator by dividing both by 2 gives:

$$\frac{\delta y}{\delta x} = \cos\left(x + \frac{\delta x}{2}\right) \frac{\sin \dfrac{\delta x}{2}}{\dfrac{\delta x}{2}}$$

In this expression, the second factor has the form $\frac{\sin\theta}{\theta}$ with $\theta = \frac{\delta x}{2}$. In Section 2.4, it was shown that $\lim\limits_{\theta \to 0}\frac{\theta}{\sin\theta} \to 1$

It follows that:

$$\lim_{\delta x \to 0} \frac{\sin \dfrac{\delta x}{2}}{\dfrac{\delta x}{2}} = 1$$

Taking the limit as $\delta x \to 0$ in the expression for $\frac{\delta y}{\delta x}$ gives:

$$\lim_{\delta x \to 0} \frac{\delta y}{\delta x} = \lim_{\delta x \to 0}\left(\cos\left(x + \frac{\delta x}{2}\right) \frac{\sin \dfrac{\delta x}{2}}{\dfrac{\delta x}{2}} \right)$$

so

$$\frac{dy}{dx} = \cos x$$

Nugget – Picturing the gradient function of $y = \sin x$

Here would be a good opportunity to step back from this last result (i.e. that the derivative of $\sin x$ is $\cos x$) and think about what it might mean graphically. Opposite is the graph of the function $y = \sin x$ over the range of values 0 to 2π (note that the angles are measured in radians, rather than degrees, so 360° corresponds to just over 6 radians).

Take each of the points A to E in turn, make a very rough estimate of the gradient at that point and plot this gradient value on the same graph. For example, at Point A (the origin), the x-coordinate is 0 and the slope is roughly 45°, giving a gradient of about 1. So you plot the first point at roughly (0, 1). You should find that the points take the general pattern of the cosine function. The graph of y = cos x is shown below to help you check that this is the case.

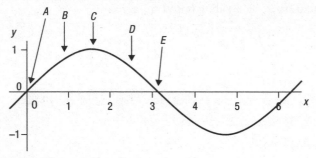

Figure 8.1a The graph of y = sin x

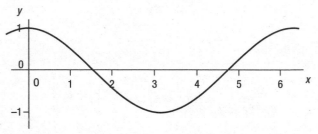

Figure 8.1b The graph of y = cos x

8.3 Differentiating cos x

Using the notation and method employed with sin x you obtain:

$$\delta y = \cos(x + \delta x) - \cos x$$

Using the trigonometric formula:

$$\cos P - \cos Q = -2\sin\frac{P+Q}{2}\sin\frac{P-Q}{2}$$

$$\delta y = -2\sin\left(x + \frac{\delta x}{2}\right)\sin\frac{\delta x}{2}$$

Dividing by δx gives

$$\frac{\delta y}{\delta x} = -2\sin\left(x + \frac{\delta x}{2}\right)\frac{\sin\frac{\delta x}{2}}{\delta x}$$

$$= -\sin\left(x + \frac{\delta x}{2}\right)\frac{\sin\frac{\delta x}{2}}{\frac{\delta x}{2}} \qquad \text{(as in Section 8.2)}$$

Therefore

$$\lim_{\delta x \to 0}\frac{\delta y}{\delta x} = \lim_{\delta x \to 0}\left(-\sin\left(x + \frac{\delta x}{2}\right)\frac{\sin\frac{\delta x}{2}}{\frac{\delta x}{2}}\right)$$

giving

$$\frac{dy}{dx} = -\sin x$$

8.4 Differentiating tan$x$

You can find the derivative of $\tan x$ by using the quotient rule and the derivatives of $\sin x$ and $\cos x$ in Sections 8.2 and 8.3.

As $\tan x = \dfrac{\sin x}{\cos x}$

$$\frac{dy}{dx} = \frac{(\cos x \times \cos x) - \{\sin x \times (-\sin x)\}}{\cos^2 x}$$

$$= \frac{\cos^2 x + \sin^2 x}{\cos^2 x}$$

$$= \frac{1}{\cos^2 x}$$

Therefore $\quad \dfrac{dy}{dx} = \sec^2 x$

You could prove this result from first principles using the same methods as for $\sin x$ and $\cos x$ using appropriate trigonometric formulae, but it would be a longer process.

Nugget – The six trig functions

In case you are rather rusty with your trigonometric functions, here is a quick reminder.

$\sin A = \frac{a}{c}$ $\cos A = \frac{b}{c}$ $\tan A = \frac{a}{b}$

$\operatorname{cosec} A = \frac{1}{\sin A} = \frac{c}{a}$ $\sec A = \frac{1}{\cos A} = \frac{c}{b}$ $\cot A = \frac{1}{\tan A} = \frac{b}{a}$

Also, $\tan A = \frac{\sin A}{\cos A}$

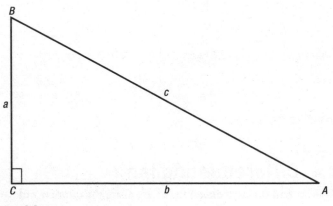

Figure 8.2

8.5 Differentiating sec x, cosec x, cot x

You can find the derivatives of these functions by expressing them as the reciprocals of $\cos x$, $\sin x$ and $\tan x$ and using the quotient rule for differentiation.

DIFFERENTIATING $y = \sec x$

If $y = \sec x$ then $y = \frac{1}{\cos x}$

Therefore

$$\frac{dy}{dx} = \frac{\cos x \times 0 - (-\sin x) \times 1}{\cos^2 x}$$

$$= \frac{\sin x}{\cos^2 x}$$

$$= \frac{1}{\cos x} \times \frac{\sin x}{\cos x}$$

Therefore $$\frac{dy}{dx} = \sec x \tan x$$

DIFFERENTIATING $y = \operatorname{cosec} x$

If $y = \operatorname{cosec} x$ then $y = \frac{1}{\sin x}$

Therefore
$$\frac{dy}{dx} = \frac{(-\cos x)}{\sin^2 x}$$
$$= -\frac{\cos x}{\sin^2 x}$$
$$= -\frac{1}{\sin x} \times \frac{\cos x}{\sin x}$$

Therefore $$\frac{dy}{dx} = -\operatorname{cosec} x \cot x$$

DIFFERENTIATING $y = \cot x$

If $y = \cot x$ then $y = \frac{\cos x}{\sin x}$

Therefore
$$\frac{dy}{dx} = \frac{\sin x \times (-\sin x) - \cos x \times \cos x}{\sin^2 x}$$
$$= \frac{-\sin^2 x - \cos^2 x}{\sin^2 x}$$
$$= -\frac{1}{\sin^2 x}$$

Therefore $$\frac{dy}{dx} = -\operatorname{cosec}^2 x$$

8.6 Summary of results

Function	Derivative
$\sin x$	$\cos x$
$\cos x$	$-\sin x$
$\tan x$	$\sec^2 x$
$\operatorname{cosec} x$	$-\operatorname{cosec} x \cot x$
$\sec x$	$\sec x \tan x$
$\cot x$	$-\operatorname{cosec}^2 x$

8.7 Differentiating trigonometric functions

Differentiating trigonometric functions often requires the 'function of a function' rule of Section 6.4. A common form involves a multiple of x, for example ax. This is a function of a function, with derivative a. Hence this must appear as a factor of the derivative.

Thus if:
$$y = \sin ax \qquad \frac{dy}{dx} = a\cos ax$$

$$y = \cos ax \qquad \frac{dy}{dx} = -a\sin ax$$

$$y = \tan ax \qquad \frac{dy}{dx} = a\sec^2 ax$$

and similarly for their reciprocals.

Thus
$$\frac{d}{dx}\sin 2x = 2\cos 2x$$

$$\frac{d}{dx}\cos\frac{x}{2} = -\frac{1}{2}\sin\frac{x}{2}$$

$$\frac{d}{dx}\tan\frac{ax}{b} = \frac{a}{b}\sec^2\frac{ax}{b}$$

Here are some more complicated forms.

$$y = \sin(ax+b) \qquad \frac{dy}{dx} = a\cos(ax+b)$$

$$y = \sin(\pi+nx) \qquad \frac{dy}{dx} = n\cos(\pi+nx)$$

$$y = \tan(1-x) \qquad \frac{dy}{dx} = -\sec^2(1-x)$$

$$y = \sin\left(\frac{1}{x^2}\right) \qquad \frac{dy}{dx} = -\frac{2}{x^3}\cos\left(\frac{1}{x^2}\right)$$

Example 8.7.1

Differentiate $y = \sin^2 x$

First rewrite $y = \sin^2 x$ as $y = (\sin x)^2$

$$\frac{dy}{dx} = 2\sin x \times \frac{d(\sin x)}{dx}$$
$$= 2\sin x \cos x = \sin 2x$$

Example 8.7.2

Differentiate $y = \sin\sqrt{x}$

First rewrite this as $y = \sin x^{\frac{1}{2}}$

Then
$$\frac{dy}{dx} = \cos x^{\frac{1}{2}} \times \frac{1}{2} x^{-\frac{1}{2}}$$
$$= \frac{1}{2}\cos\sqrt{x} \times \frac{1}{\sqrt{x}} = \frac{\cos\sqrt{x}}{2\sqrt{x}}$$

Example 8.7.3

Differentiate $y = \sqrt{\sin x}$

First rewrite this as $y = (\sin x)^{\frac{1}{2}}$

Then
$$\frac{dy}{dx} = \frac{1}{2}(\sin x)^{-\frac{1}{2}} \times \frac{d(\sin x)}{dx}$$
$$= \frac{1}{2}\frac{1}{\sqrt{\sin x}} \times \cos x = \frac{\cos x}{2\sqrt{\sin x}}$$

Example 8.7.4

Differentiate $y = \sin^2 x^2$

First rewrite this as $y = (\sin x^2)^2$

Then
$$\frac{dy}{dx} = 2\sin x^2 \times \cos x^2 \times 2x$$
$$= 4x \sin x^2 \cos x^2$$

Exercise 8.1

Differentiate the following.

1 $3\sin x$

2 $\sin 3x$

3 $\cos \frac{1}{2}x$

4 $\tan \frac{1}{3}x$

5 $\sec 0.6x$

6 $\csc \frac{1}{6}x$

7 $\sin 2x + 2\cos 2x$

8 $\sin 3x - \cos 3x$

9 $\sec x + \tan x$

10 $\sin 4x + \cos 5x$

11 $\cos \frac{1}{2}\theta + \sin \frac{1}{4}\theta$

12 $\sin\left(2x + \frac{1}{2}\pi\right)$

13 $\cos(3\pi - x)$

14 $\csc\left(a - \frac{1}{2}x\right)$

15 $\sin^3 x$

16 $\sin(x^3)$

17 $\cos^3(2x)$

18 $\sec x^2$

19 $\tan \sqrt{1-x}$

20 $2\tan\left(\frac{1}{2}x\right)$

21 $a(1 - \cos x)$

22 $a \sin nx + b \cos nx$

23 $\cos\left(2x + \frac{1}{2}\pi\right)$

24 $\tan 2x - \tan^2 x$

25 $x^2 + 3\sin \frac{1}{2}x$

26 $x \sin x$

27 $\cos \frac{a}{x}$

28 $\dfrac{x}{\sin x}$

29 $x \tan x$

30 $\sin 2x + \sin(2x)^2$

31 $\dfrac{\tan x}{x}$

32 $\dfrac{x}{\tan x}$

33 $\cos^3(x^2)$

34 $x^2 \tan x$

35 $\cot(5x + 1)$

36 $\cot^2 3x$

37 $\sqrt{\cos x}$

38 $\sin 2x \cos 2x$

39 $\cos^2 x + \sin^2 x$

40 $\sin^2 x - \cos^2 x$

41 $\dfrac{1}{1 + \cos x}$

42 $\dfrac{1 - \cos x}{1 + \cos x}$

43 $\dfrac{\sqrt{x}}{\sin x}$

44 $\dfrac{1}{1 - \tan x}$

45 $\dfrac{x^2}{\cos 2x}$

46 $\dfrac{\tan x - 1}{\sec x}$ **47** $x\sqrt{\sin x}$ **48** $\dfrac{\sin^2 x}{1+\sin x}$

49 $x^2 \cos 2x$ **50** $\sec^2 x \operatorname{cosec} x$

8.8 Successive derivatives

Let $y = \sin x$ so that
$$\frac{dy}{dx} = \cos x$$
$$\frac{d^2 y}{dx^2} = -\sin x$$
$$\frac{d^3 y}{dx^3} = -\cos x$$
$$\frac{d^4 y}{dx^4} = \sin x$$

These derivatives will now repeat in identical sets of four.

Using the trigonometric formula $\cos x = \sin\left(x + \tfrac{1}{2}\pi\right)$, you can rewrite these equations in the form:

$$\frac{dy}{dx} = \cos x = \sin\left(x + \tfrac{1}{2}\pi\right)$$
$$\frac{d^2 y}{dx^2} = \frac{d}{dx}\left\{\sin\left(x + \tfrac{1}{2}\pi\right)\right\} = \cos\left(x + \tfrac{1}{2}\pi\right) = \sin\left(x + \tfrac{2}{2}\pi\right)$$
$$\frac{d^3 y}{dx^3} = \frac{d}{dx}\left\{\sin\left(x + \tfrac{2}{2}\pi\right)\right\} = \cos\left(x + \tfrac{2}{2}\pi\right) = \sin\left(x + \tfrac{3}{2}\pi\right)$$

You can continue indefinitely, adding $\tfrac{1}{2}\pi$ each time.

Thus you can deduce that $\dfrac{d^n y}{dx^n} = \sin\left(x + \dfrac{n\pi}{2}\right)$

You can obtain successive derivatives of $\cos x$ in a similar way. The successive derivatives of $\tan x$, $\sec x$ and the other trigonometric functions become very complicated after a few differentiations, and cannot be expressed using simple formulae.

8.9 Graphs of the trigonometric functions

Example 8.9.1 The graphs of sin x and cos x

When $y = \sin x$, $\frac{dy}{dx} = \cos x$ and $\frac{d^2y}{dx^2} = -\sin x$. These graphs are drawn in Figure 8.3. The thickest graph is $y = \sin x$, the dashed graph is $\frac{dy}{dx}$ and the thin graph is $\frac{d^2y}{dx^2}$

Since $\sin x = \sin(x + 2\pi)$, the part of the curve between $x = 0$ and $x = 2\pi$ will be repeated for intervals of 2π as x increases. There will be similar sections for negative angles.

Thus the part of the graph of $y = \sin x$ between 0 and 2π will be repeated an infinite number of times between $-\infty$ and $+\infty$, the whole forming one continuous curve.

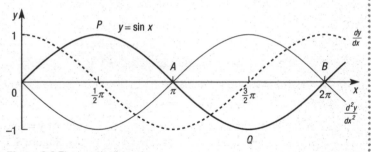

Figure 8.3 The graph of $y = \sin x$

The function $\sin x$ is an example of a periodic function. A **periodic function** is a function which repeats itself at regular intervals, the smallest interval being called the **period** of the function. Thus 2π is the period of the function $\sin x$.

The following characteristics of the graph of $y = \sin x$ illustrate much of the work of the preceding chapter.

Curvature The graph of $y = \sin x$ between 0 and 2π provides examples of the four types of curvature illustrated in Figures 7.2a and 7.2b and 7.3a and 7.3b. The graph of $\frac{dy}{dx}$ illustrates the connection between these different types of curvature and the sign of the derivative, discussed in Section 7.1.

Turning points The curve between 0 and 2π shows that between these two values of x there are two turning points, at P and Q, the values being -1 and 1.

At P, when $x = \frac{\pi}{2}$, $\frac{dy}{dx} = 0$ and $\frac{d^2y}{dx^2}$ is negative, showing that P is a local maximum. At Q, when $x = \frac{3\pi}{2}$, $\frac{dy}{dx} = 0$ and $\frac{d^2y}{dx^2}$ is positive, showing that Q is a local minimum.

This is true for any part of the curve of width 2π. Consequently throughout the curve from $-\infty$ and $+\infty$ there is an infinite sequence of turning points, alternately local maxima and local minima.

Points of inflexion There are two points of inflexion on the section of the curve between A and B.

At A the curve changes from concave down to concave up:

$$\frac{dy}{dx} \text{ has a minimum value of } -1$$

$$\frac{d^2y}{dx^2} = 0 \text{ and is changing from negative to positive}$$

At B this is reversed. The curve changes from concave up to concave down:

$$\frac{dy}{dx} \text{ has a maximum value of } 1$$

$$\frac{d^2y}{dx^2} = 0 \text{ and is changing from from positive to negative. There}$$
is also a point of inflexion at the origin.

Note that the graph of $y = \cos x$ is the same as the graph of $y = \sin x$ moved $\frac{1}{2}\pi$ to the left along the x-axis. The graph of $\frac{dy}{dx}$ in Figure 8.3 shows its shape and position. Consequently all the results for $y = \sin x$ are repeated for $y = \cos x$ with the angle reduced by $\frac{1}{2}\pi$.

Example 8.9.2 The graphs of tan x and cot x

When:

$y = \tan x$

$\dfrac{dy}{dx} = \sec^2 x$

$\dfrac{d^2y}{dx^2} = 2\sec^2 x \tan x$

When:

$y = \cot x$

$\dfrac{dy}{dx} = -\operatorname{cosec}^2 x$

$\dfrac{d^2y}{dx^2} = 2\operatorname{cosec}^2 x \cot x$

The graphs of tan x and of its derivative sec^2 x are shown in Figure 8.4, the latter curve being dashed.

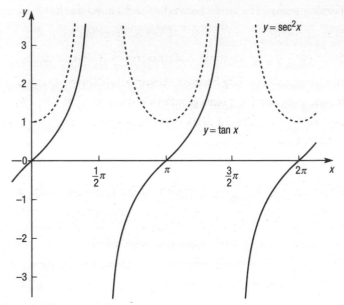

Figure 8.4 The graph of $y = \tan x$

Note the following characteristics of the graph of $y = \tan x$.

The curve is discontinuous. When $x \to \frac{1}{2}\pi$ $\tan x \to \infty$. On passing through $\frac{1}{2}\pi$ an infinitely small increase in x results in the angle being in the second quadrant. Its tangent is therefore negative, while still numerically infinitely great. With this infinitely small increase in x, $\tan x$ changes from $+\infty$ to $-\infty$.

The graph of the function $y = \tan x$ is therefore discontinuous. Similar changes occur when $x = \frac{3}{2}\pi, \frac{5}{2}\pi$, and so on. You can see this in Figure 8.4.

The graph of $\tan x$ is periodic and the period is π.

The function is always increasing. This is indicated by the fact that the derivative $\frac{dy}{dx} = \sec^2 x$ is always positive.

There is a point of inflexion when $x = \pi$. The curve is changing from concave down to concave up, the derivative, $\sec^2 x$, is a minimum, and its value is $+1$.

Consequently the curve crosses the x-axis at an angle of $\frac{1}{4}\pi$.

Similar points occur if $x = 0$ and any integral multiple of π.

Since $\cot x = \frac{1}{\tan x}$ its graph is the inversion of that of $\tan x$.

It is always decreasing because $-\mathrm{cosec}^2\, x$ is always negative; it is periodic and has points of inflexion when $x = \frac{1}{2}\pi, \frac{3}{2}\pi, \frac{5}{2}\pi$, and so on. You should draw it on your calculator or spreadsheet.

Example 8.9.3 The graphs of cosec x and sec x

Turning points on these curves may be deduced from those of their reciprocals. When $\sin x$ is a maximum, $\mathrm{cosec}\, x$ is a minimum; consequently the graphs are periodic, and maximum and minimum values occur alternately.

If
$$y = \mathrm{cosec}\, x$$

then
$$\frac{dy}{dx} = -\mathrm{cosec}\, x \cot x$$

When $x = \frac{1}{2}\pi$, $-\mathrm{cosec}\, x = -1$, $\cot x = 0$, so $\frac{dy}{dx} = 0$

The value of $\frac{d^2 y}{dx^2}$ will be found to be positive. Hence there is a minimum value when $x = \frac{1}{2}\pi$

Both curves are discontinuous and periodic.

Example 8.9.4

Find the turning points on the graph of $y = \sin x + \cos x$

If
$$y = \sin x + \cos x$$

then
$$\frac{dy}{dx} = \cos x - \sin x$$

For turning points
$$\frac{dy}{dx} = 0$$

Putting $\cos x - \sin x = 0$ gives $\cos x = \sin x$ or $\tan x = 1$

Therefore
$$x = \frac{1}{4}\pi$$

But this is the smallest of a series of angles with tangent equal to $+1$. All the angles are included in the general formula:

$$x = n\pi + \frac{1}{4}\pi$$

Therefore the angles for which there are turning points are

$$\frac{1}{4}\pi, \frac{5}{4}\pi, \frac{9}{4}\pi, \ldots$$

Also
$$\frac{d^2y}{dx^2} = -\sin x - \cos x$$

This is negative at $\frac{1}{4}\pi, \frac{9}{4}\pi, \ldots$ and positive when $x = \frac{5}{4}\pi, \frac{13}{4}\pi, \ldots$

Therefore the curve is periodic, and local maximum and minimum values occur alternately.

Maxima occur when
$$x = \frac{1}{4}\pi, \frac{9}{4}\pi, \ldots$$

Minima occur when
$$x = \frac{5}{4}\pi, \frac{13}{4}\pi, \ldots$$

The maximum value is

$$\sin\frac{1}{4}\pi + \cos\frac{1}{4}\pi = \frac{1}{\sqrt{2}} + \frac{1}{\sqrt{2}} = \frac{2}{\sqrt{2}} = \sqrt{2}$$

Similarly the minimum value is $-\sqrt{2}$

The curve is shown as the heavy line in Figure 8.5. The dotted curves are $\sin x$ and $\cos x$. P is the maximum point and Q the minimum. A is a point of inflexion.

The curve is a simple example of a harmonic curve or a wave diagram, which are used extensively in electrical engineering.

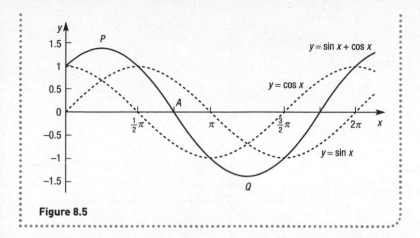

Figure 8.5

8.10 Inverse trigonometric functions

When you write $y = \sin x$ the sine is expressed as a function of the angle x. When x varies, the sine varies too, that is, the angle is the independent variable and y or $\sin x$ is the dependent variable.

But you may need to reverse this relation to express the angle as a function of the sine. Thus we express the fact that when $\sin x$ changes, the angle changes in consequence. The value of y or $\sin x$ is now the independent variable and the angle is the dependent variable.

This relation, as you know from trigonometry, is expressed by the form:

$$y = \sin^{-1} x$$

which means, y is the angle of which x is the sine. From this you can write down the direct function relation, that is:

$$x = \sin y$$

You need to take some care because a function must only have one value of y for a given value of x. Moreover, if you press the $\sin^{-1}$ key on the calculator, it gives you a single value of y, between $-\frac{1}{2}\pi$ and $\frac{1}{2}\pi$. The graph of $y = \sin^{-1} x$ is shown in Figure 8.6.

You must note that the -1 is not an index, but a part of the symbol $\sin^{-1}$, which denotes the inverse function. All the other trigonometric functions can similarly be expressed in inverse form, with similar restrictions on the values of y.

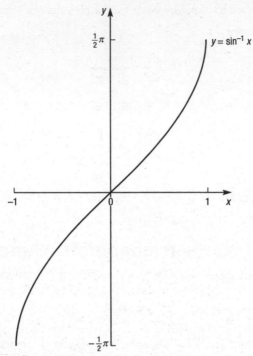

Figure 8.6

8.11 Differentiating $\sin^{-1}x$ and $\cos^{-1}x$

Let $\qquad\qquad\qquad\qquad y = \sin^{-1}x$

Then, as shown, $\qquad\qquad x = \sin y$

Differentiating x with respect to y

$$\frac{dx}{dy} = \cos y$$

Therefore $\qquad\qquad \dfrac{dy}{dx} = \dfrac{1}{\dfrac{dx}{dy}} = \dfrac{1}{\cos y}$

From the relation $\sin^2 y + \cos^2 y = 1$

$$\cos y = \pm\sqrt{1 - \sin^2 y} = \pm\sqrt{1 - x^2}$$

But as $-\frac{1}{2}\pi < y < \frac{1}{2}\pi$, and $\cos y > 0$ for all values of y in this interval, $\cos y = +\sqrt{1-x^2}$

Hence

$$\frac{dy}{dx} = \frac{1}{\sqrt{1-x^2}}$$

Similarly, if

$$y = \cos^{-1} x$$

then

$$\frac{dy}{dx} = -\frac{1}{\sqrt{1-x^2}}$$

8.12 Differentiating $\tan^{-1} x$ and $\cot^{-1} x$

Let

$$y = \tan^{-1} x$$

Then

$$x = \tan y \text{ with } -\frac{1}{2}\pi < y < \frac{1}{2}\pi$$

Differentiating with respect to y

$$\frac{dx}{dy} = \sec^2 y$$

and

$$\frac{dy}{dx} = \frac{1}{\sec^2 y} = \frac{1}{1+\tan^2 y} = \frac{1}{1+x^2}$$

Therefore

$$\frac{dy}{dx} = \frac{1}{1+x^2}$$

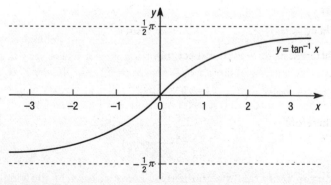

Figure 8.7 Graph of $y = \tan^{-1} x$

The following points are shown by the graph of $\tan^{-1} x$ in Figure 8.7.

▶ $\frac{dy}{dx}$ is always positive so y is always increasing.

▶ $\frac{dy}{dx}$ is never zero, so there are no turning points.

You should draw the graph of $y = \cot^{-1} x$, that is, $y = \tan^{-1}\frac{1}{x}$, on your calculator or spreadsheet. You will see that $y = \cot^{-1} x$ is not defined for $x = 0$. If you use a method similar to that for differentiating $\tan^{-1} x$ on $y = \cot^{-1} x$ you can show that

$$\frac{dy}{dx} = -\frac{1}{1+x^2} \text{ provided } x \neq 0$$

Thus the gradient is always negative, and there are no turning points.

8.13 Differentiating $\sec^{-1} x$ and $\operatorname{cosec}^{-1} x$

Let
$$y = \sec^{-1} x$$

Then
$$x = \sec y, \frac{dx}{dy} = \sec y \tan y$$

so
$$\frac{dy}{dx} = \frac{1}{\sec y \tan y}$$

But
$$\tan y = \pm\sqrt{\sec^2 y - 1} = \pm\sqrt{x^2 - 1}$$

To decide which sign to use, first note that y is not defined when $x = 0$. In the interval $0 < y < \pi$, omitting $y = \frac{1}{2}\pi$ where $\sec y \tan y$ is not defined, $\sec y \tan y > 0$. Both are positive in the interval $0 < y < \frac{1}{2}\pi$ and both are negative in $\frac{1}{2}\pi < y < \pi$. Therefore $\frac{dy}{dx} = \sec y \tan y > 0$ for all $0 < y < \pi$ omitting $y = \frac{1}{2}\pi$. Therefore

$$\frac{dy}{dx} = \left| \frac{1}{x\sqrt{x^2 - 1}} \right|$$

You can verify that the gradient is always positive by drawing the graph of $y = \sec^{-1} x$, which is shown in Figure 8.8.

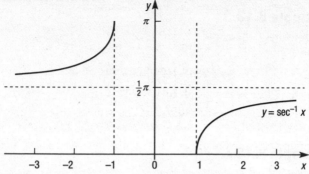

Figure 8.8 Graph of $y = \sec^{-1} x$

There are no turning points, and the graph is not defined when x lies between −1 and 1. The graph of $y = \text{cosec}^{-1} x$ has similar properties, and you would find that

$$\frac{dy}{dx} = -\left| \frac{1}{x\sqrt{x^2 - 1}} \right|$$

Example 8.13.1

Differentiate $y = \sin^{-1} \frac{x}{a}$

Using the rule for a function of a function

$$\frac{dy}{dx} = \frac{1}{\sqrt{1 - \left(x^2 / a^2 \right)}} \times \frac{1}{a}$$

$$= \frac{1}{\sqrt{a^2 - x^2}}$$

Example 8.13.2

Differentiate $y = \sin^{-1} x^2$

$$\frac{dy}{dx} = \frac{1}{\sqrt{1 - (x^2)^2}} \times \frac{d}{dx}(x^2)$$

$$= \frac{2x}{\sqrt{1 - x^4}}$$

Example 8.13.3

Differentiate $y = \tan^{-1}\frac{1}{x^2}$

$$\frac{dy}{dx} = \frac{1}{1+\left(\dfrac{1}{x^2}\right)^2} \times \frac{d}{dx}\left(\frac{1}{x^2}\right) = \frac{1}{1+\dfrac{1}{x^4}} \times \frac{-2}{x^3} = \frac{-2x}{x^4+1}$$

Example 8.13.4

Differentiate $y = x^2 \sin^{-1}(1-x)$

Using the product rule

$$\frac{dy}{dx} = 2x\sin^{-1}(1-x) + x^2 \times \frac{1}{\sqrt{1-(1-x)^2}} \times -1$$

$$= 2x\sin^{-1}(1-x) - \frac{x^2}{\sqrt{2x-x^2}}$$

8.14 Summary of results

The derivatives of the inverse trigonometric functions are collected together below for reference.

Function	Derivative
$\sin^{-1}x$	$\dfrac{1}{\sqrt{1-x^2}}$
$\cos^{-1}x$	$-\dfrac{1}{\sqrt{1-x^2}}$
$\tan^{-1}x$	$\dfrac{1}{1+x^2}$
$\cot^{-1}x$	$-\dfrac{1}{1+x^2}$
$\sec^{-1}x$	$\left|\dfrac{1}{x\sqrt{x^2-1}}\right|$
$\operatorname{cosec}^{-1}x$	$-\left|\dfrac{1}{x\sqrt{x^2-1}}\right|$

Nugget – Remembering the formulae

In mathematics there are some formulae that you can look up in a book and there are others that you really need to learn and remember because they crop up so often. It is very much up to you which formulae you think are worth remembering as it depends on how often you need them. However, here is a suggestion.

In Section 8.6 you were provided with a table summarizing the derivatives of the six trigonometric functions. Learn them and remember them all!

In Section 8.14 you are provided with a table summarizing the derivatives of the six inverse trigonometric functions. My advice here is to learn and remember the first three of these.

Exercise 8.2

In questions 1 to 8 find for which values of x, in the given interval, there are local maximum or local minimum values, and decide in each case whether the value is a maximum or a minimum.

1 $\sin 2x - x$, $0 < x < \pi$

2 $\sin^2 x \cos^2 x$, $0 < x < \pi$

3 $\sin x + \sin x \cos x$, $0 < x < \pi$

4 $\dfrac{\sin x}{1 + \tan x}$, $0 < x < \pi$

5 $2 \sin x + \cos x$, $0 < x < \pi$

6 $\sin x + \cos x$, $0 < x < \pi$

7 $2 \sin x - \sin 2x$, $0 < x < 2\pi$

8 $\sin x \sin 2x$, $-\dfrac{1}{2}\pi < x < \dfrac{1}{2}\pi$

9 What is the smallest value of x for which $2 \sin x + 3 \cos x$ is a maximum?

10 Find the smallest value of x for which $\tan^2 x - 2 \tan x$ is a local maximum or minimum.

Differentiate the following functions.

11 $\sin^{-1} 4x$ **12** $\sin^{-1} \dfrac{1}{2} x$ **13** $b\cos^{-1} \dfrac{x}{a}$

14 $\cos^{-1} \dfrac{x}{3}$ **15** $\tan^{-1} \dfrac{x}{3}$ **16** $\tan^{-1}(a-x)$

17 $\cos^{-1} 2x^2$ **18** $\sin^{-1} \sqrt{x}$ **19** $x\sin^{-1} x$

20 $\sin^{-1} \dfrac{1}{x}$

Key ideas

▶ In this chapter you were given the derivatives of the six common trigonometric functions and these are summarized in Section 8.6.

▶ You were shown the characteristic shapes of the graphs of $\sin x$, $\cos x$ and $\tan x$.

▶ The graphs of $\sin x$, $\cos x$ and $\tan x$ are *periodic*. There are several characteristics of a periodic function, the chief one being that its graph repeats itself at regular intervals.

▶ The *inverse function of a trigonometric function* such as $\sin x$ is $\sin^{-1} x$ and this means 'the angle whose sin is x'.

▶ The table in Section 8.14 summarizes the derivatives of the six inverse trigonometric functions.

9

Exponential and logarithmic functions

In this chapter you will learn:

▶ *that a study of compound interest leads to the exponential function e^x*

▶ *that the derivative of e^x is e^x*

▶ *how to differentiate the inverse of the exponential function, $\ln x$.*

9.1 Compound Interest Law of growth

Nugget – Compound interest and exponential growth

Just reflect for a moment why this chapter (which is principally on the exponential function) should have started with an explanation of how compound interest is calculated. The reason is that compound interest is an excellent practical context for understanding how **exponential growth** works. The basic idea underlying them both is that there is a fixed percentage growth each period.

For example, suppose you start with, say £500 and this increases by 4% each year. Then in successive years, the value of your investment can be summarized as follows.

Year	0	1	2	3...	t
Amount (£)	500	$500 \times (1 + \frac{4}{100})$	$500 \times (1 + \frac{4}{100})^2$	$500 \times (1 + \frac{4}{100})^3 \cdots$	$500 \times (1 + \frac{4}{100})^t$

As you will see, this formula, $A = 500 \times (1 + \frac{4}{100})^t$, is of the basic form for describing exponential change – i.e. where there is a fixed percentage change year on year.

You will probably be familiar with two methods of payment of interest on borrowed or invested money, called simple interest and compound interest. In each of them, the interest bears a fixed ratio to the magnitude of the sum of money involved, called the **principal**. But while with **simple interest** the principal remains the same from year to year, with **compound interest** the interest is added to the principal at the end of each year, over a period, and the interest for the succeeding year is calculated on the sum of the principal and interest.

Let P be the principal and r be the rate per cent, per annum.

The interest added at the end of the first year is $P \times \dfrac{r}{100}$

The total at the end of the first year is $P + P \times \dfrac{r}{100} = P\left(1 + \dfrac{r}{100}\right)$

This is the principal for the new year. Therefore, by the same working as for the first year:

the total at the end of the second year is $P\left(1 + \dfrac{r}{100}\right)^2$

the total at the end of the third year is $P\left(1 + \dfrac{r}{100}\right)^3$

the total at the end of the fourth year is $P\left(1 + \dfrac{r}{100}\right)^4$

Suppose that the interest is added at the end of each half year instead of at the end of each year, then:

the total at the end of the first half-year is $P\left(1 + \dfrac{r}{2 \times 100}\right)$

the total at the end of the first year is $P\left(1 + \dfrac{r}{2 \times 100}\right)^2$

the total at the end of the second year is $P\left(1 + \dfrac{r}{2 \times 100}\right)^4$

the total at the end of t years is $P\left(1 + \dfrac{r}{2 \times 100}\right)^{2t}$

If the interest is added four times a year:

the total at the end of the first year is $P\left(1 + \dfrac{r}{4 \times 100}\right)^4$

the total at the end of t years is $P\left(1 + \dfrac{r}{4 \times 100}\right)^{4t}$

Similarly, if the interest is added monthly, that is, 12 times a year:

the total at the end of the tth year is $P\left(1+\dfrac{r}{12\times100}\right)^{12t}$

If the interest is added m times a year:

the total at the end of t years is $P\left(1+\dfrac{r}{100m}\right)^{mt}$

In this result let $\dfrac{r}{100m}=\dfrac{1}{n}$, so that $m=\dfrac{nr}{100}$

Then you can write the total after t years as:

$$P\left(1+\frac{1}{n}\right)^{\frac{nrt}{100}}=P\left\{\left(1+\frac{1}{n}\right)^{n}\right\}^{\frac{rt}{100}}$$

Now, suppose that n becomes indefinitely large, that is, the interest is added on at indefinitely small intervals, so that the growth of the principal may be regarded as continuous.

Then the total reached after t years will be the limit of

$$P\left\{\left(1+\frac{1}{n}\right)^{n}\right\}^{\frac{rt}{100}}$$

as n becomes infinitely large.

To find this you need to find the limit of $\left(1+\dfrac{1}{n}\right)^{n}$ as $n\rightarrow\infty$, that is, the total reached after t years is

$$P\left\{\lim_{n\to\infty}\left(1+\frac{1}{n}\right)^{n}\right\}^{\frac{rt}{100}}$$

It becomes necessary, therefore, to find the value of

$$\lim_{n\to\infty}\left(1+\frac{1}{n}\right)^{n}$$

9.2 The value of $\lim\limits_{n \to \infty} \left(1 + \dfrac{1}{n}\right)^n$

Try finding the value of $\lim\limits_{n \to \infty} \left(1 + \dfrac{1}{n}\right)^n$ by using your calculator or

spreadsheet for increasingly large values of n. The values achieved, using a spreadsheet, are shown in the following table.

Value of n	Value of $\left(1 + \dfrac{1}{n}\right)^n$
1	2
10	2.593742
100	2.704814
1000	2.716924
10 000	2.718146
100 000	2.718268
1 000 000	2.718280
10 000 000	2.718282
100 000 000	2.718282
1 000 000 000	2.718282

Assume now that $\lim\limits_{n \to \infty} \left(1 + \dfrac{1}{n}\right)^n = 2.718\ 282$ correct to six decimal places.

This value is always denoted by the letter e.

Thus $\qquad e = \lim\limits_{n \to \infty} \left(1 + \dfrac{1}{n}\right)^n$

Using this result in the expression

$$A = P \left\{ \lim\limits_{n \to \infty} \left(1 + \dfrac{1}{n}\right)^n \right\}^{\frac{rt}{100}}$$

given at the end of Section 9.1 (for the total when compound interest is continuously added) you find that

$$A = Pe^{\frac{rt}{100}}$$

Let
$$\frac{rt}{100} = x$$

Then you can write
$$A = Pe^x$$

Definition: The function e^x is called an **exponential function** because the index or exponent is the variable part of the function.

9.3 The Compound Interest Law

The fundamental principle employed in arriving at the above result is that the growth of the principal is continuous in time and does not take place by sudden increases at regular intervals. In practice, compound interest is added at definite intervals of time, but the phenomenon of continuous growth is a natural law of organic growth and change. In many physical, chemical, electrical and engineering processes, the mathematical expressions of them involve functions in which the variation is proportional to the functions themselves. In such cases the exponential function will be involved, and as the fundamental principle is that which entered into the compound interest investigations above, this law of growth was called by Lord Kelvin the Compound Interest Law.

9.4 Differentiating e^x

Let $y = e^x$ Let δx be an increase in x Let δy be the corresponding increase in y

Then
$$y + \delta y = e^{x+\delta x}$$

and
$$\delta y = e^{x+\delta x} - e^x$$

Dividing by δx gives $\dfrac{\delta y}{\delta x} = \dfrac{e^{x+\delta x} - e^x}{\delta x}$

The next step is to find the limit of the right-hand side as $\delta x \to 0$. This requires some manipulation using the rules of indices.

Thus
$$\frac{\delta y}{\delta x} = \frac{e^{x+\delta x} - e^x}{\delta x}$$
$$= \frac{e^x(e^{\delta x} - 1)}{\delta x} = e^x \times \frac{(e^{\delta x} - 1)}{\delta x}$$

and
$$\lim_{\delta x \to 0} \frac{\delta y}{\delta x} = \lim_{\delta x \to 0} \left\{ e^x \frac{(e^{\delta x} - 1)}{\delta x} \right\}$$
$$= e^x \times \lim_{\delta x \to 0} \left\{ \frac{(e^{\delta x} - 1)}{\delta x} \right\}$$

You can get information about the limit $\lim\limits_{\delta x \to 0} \left\{ \dfrac{(e^{\delta x} - 1)}{\delta x} \right\}$ by using a spreadsheet or calculator. Here are some results.

Value of δx	Value of $\left\{ \dfrac{(e^{\delta x} - 1)}{\delta x} \right\}$
0.1	1.051 710
0.01	1.005 017
0.001	1.000 500
0.000 1	1.000 050
0.000 01	1.000 005
0.000 001	1.000 001
0.000 000 1	1.000 000

It therefore appears that:

$$\lim_{\delta x \to 0} \left\{ \frac{(e^{\delta x} - 1)}{\delta x} \right\} = 1$$

Using this result in the expression for $\lim_{\delta x \to 0} \dfrac{\delta y}{\delta x}$ you find that:

$$\frac{dy}{dx} = \lim_{n \to \infty} \frac{\delta y}{\delta x} = e^x \times \lim_{\delta x \to 0} \left\{ \frac{(e^{\delta x} - 1)}{\delta x} \right\}$$

$$= e^x$$

Therefore, if $y = e^x$

$$\frac{dy}{dx} = e^x$$

Note: The results $e = \lim_{n \to \infty} \left(1 + \dfrac{1}{n} \right)^n$ and $\lim_{\delta x \to 0} \left\{ \dfrac{(e^{\delta x} - 1)}{\delta x} \right\} = 1$ are not independent.

Let $n = \dfrac{1}{\delta x}$ and suppose that n is large so that δx is small

Then for large n

$$e \approx \left(1 + \frac{1}{n} \right)^n = (1 + \delta x)^{\frac{1}{\delta x}}$$

Raising both sides to the power δx

$$e^{\delta x} \approx 1 + \delta x$$

so $\qquad\qquad e^{\delta x} - 1 \approx \delta x$

and $\qquad\qquad \dfrac{e^{\delta x} - 1}{\delta x} \approx 1$

Therefore it is not unreasonable that:

$$\lim_{\delta x \to 0}\left\{\frac{(e^{\delta x}-1)}{\delta x}\right\}=1$$

The property of e^x, that its derivative is equal to itself, is possessed by no other function of x.

Using the function of a function rule:

if $\qquad y = e^{-x} \qquad \dfrac{dy}{dx} = e^{-x} \times \dfrac{d}{dx}(-x) = -e^{-x}$

if $\qquad y = e^{ax} \qquad \dfrac{dy}{dx} = e^{ax} \times \dfrac{d}{dx}(ax) = ae^{ax}$

if $\qquad y = e^{-ax} \qquad \dfrac{dy}{dx} = -ae^{-ax}$

Nugget – If you learn nothing else...!

If you learn nothing else from this chapter, do make sure that you remember this highly memorable fact, namely that the derivative of e^x is itself, e^x. In a way, this result is one of the little miracles of mathematics and is one of the reasons that this number e is so much loved by mathematicians.

9.5 The exponential curve

GRAPH $y = e^x$

If $y = e^x$ then

$$\frac{dy}{dx} = e^x$$

$$\frac{d^2y}{dx^2} = e^x$$

Since e^x is always positive, $\frac{dy}{dx}$ is always positive, and the graph of the function e^x must be positive and always increasing. Therefore it has no turning points.

Since $\frac{d^2y}{dx^2} = e^x$ and $e^x \neq 0$ for any value of x there is no point of inflexion.

GRAPH $y = e^{-x}$

If $y = e^{-x}$ then

$$\frac{dy}{dx} = -e^{-x}$$

$$\frac{d^2y}{dx^2} = e^{-x}$$

Applying the same reasoning as for e^x above, $\frac{dy}{dx}$ is always negative, and the graph is always decreasing. There are no turning points and no points of inflexion. The two graphs are shown in Figure 9.2.

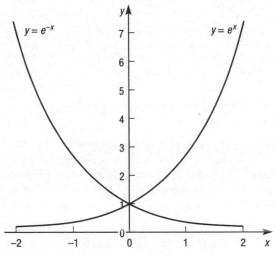

Figure 9.2 The graphs of $y = e^x$ and $y = e^{-x}$

The graph of e^{-x} shows a law of decrease common in chemical and physical processes, representing a **decay** law, the amount of decay being proportional to the magnitude of that which is diminishing at any instant. The loss of temperature in a cooling body is an example.

9.6 Natural logarithms

In Section 9.1 this formula was derived:

$$A = Pe^{\frac{rt}{100}}$$

This may be written

$$\frac{A}{P} = e^{\frac{rt}{100}}$$

Letting $\dfrac{rt}{100} = x$, this becomes $\dfrac{A}{P} = e^x$

In this form you can see x is the logarithm of $\frac{A}{P}$ to base e. In many similar examples e arises naturally as the base of a system of logarithms. These logarithms are now called **natural logarithms**.

When logarithms were first discovered by Lord Napier in 1614, the base of his system involved e. Hence these logarithms are also called Napierian logarithms. In addition they are called hyperbolic logarithms, from their association with the hyperbola. The introduction of 10 as a base was subsequently made by a mathematician called Briggs, who saw that logarithms of this type would be valuable in calculations.

In subsequent work in this book, unless it is stated to the contrary, the logarithms employed will be those to base e. The logarithm of x to the base e will be denoted by $\ln x$.

9.7 Differentiating ln x

You could find the derivative of $\ln x$ by using first principles, but it is easier to use the following indirect method based on inverse functions.

Let

$$y = \ln x$$

Then

$$x = e^y$$

and

$$\frac{dx}{dy} = e^y$$

so

$$\frac{dy}{dx} = \frac{1}{e^y} = \frac{1}{x}$$

If the logarithm involved a different base, say a where $a > 0$, then you can change it to a logarithm to base e by the following method.

Let

$$y = \log_a x$$

Then

$$x = a^y$$

Taking logarithms to the base e:

$$\ln x = \ln a^y = y \ln a$$

so

$$\frac{dy}{dx} = \frac{1}{x \ln a}$$

As a special case, if $\qquad y = \log_{10} x$

then

$$\frac{dy}{dx} = \frac{1}{x \ln 10}$$

9.8 Differentiating general exponential functions

The function e^x is a special case of the general exponential function a^x where a is any positive number.

Let $\qquad\qquad y = a^x$

Then $\qquad\qquad \ln y = x \ln a$

so $\qquad\qquad y = e^{x \ln a}$

Therefore $\qquad \dfrac{dy}{dx} = \ln a \times e^{x \ln a}$

$$\qquad\qquad\qquad = a^x \ln a$$

As a special case, if $\qquad y = 10^x$

then $\qquad\qquad \dfrac{dy}{dx} = 10^x \ln 10$

9.9 Summary of formulae

The work done on exponential and logarithm functions so far can be summarized in the following table. The results come up frequently, so you need to be familiar with them.

Function	Derivative
e^x	e^x
e^{-x}	$-e^{-x}$
a^x	$a^x \ln a$
$\ln x$	$\frac{1}{x}$

9.10 Worked examples

Example 9.10.1

Differentiate $y = e^{3x^2}$

Employing the function of a function rule, if

$$y = e^{3x^2}$$

then

$$\frac{dy}{dx} = e^{3x^2} \times \frac{d}{dx}\left(3x^2\right)$$

$$= 6xe^{3x^2}$$

Example 9.10.2

Differentiate $y = \ln x^2$

$y = \ln x^2$

$$\frac{dy}{dx} = \frac{1}{x^2} \times \frac{d}{dx}\left(x^2\right)$$

$$= \frac{1}{x^2} \times 2x = \frac{2}{x}$$

The same result can be obtained by noting that $\ln x^2 = 2\ln x$.

Example 9.10.3

Differentiate $\ln \dfrac{x^2}{\sqrt{x^2-1}}$

This may be written as $\ln x^2 - \ln \sqrt{x^2-1}$, or, manipulating it further,

$$2\ln x - \tfrac{1}{2}\ln(x^2 - 1)$$

Therefore

$$\frac{dy}{dx} = \frac{2}{x} - \frac{1}{2} \times \frac{1}{x^2-1} \times \frac{d}{dx}\left(x^2 - 1\right)$$

$$= \frac{2}{x} - \frac{x}{x^2-1}$$

Example 9.10.4

Differentiate $y = e^{-ax} \sin(bx + c)$

This type of expression is important in many electrical and physical problems, such as, for example, the decay of the swing of a pendulum in a resisting medium.

Let
$$y = e^{-ax} \sin(bx + c)$$

Then
$$\frac{dy}{dx} = -ae^{-ax} \times \sin(bx + c) + e^{-ax} \times b \cos(bx + c)$$
$$= -ae^{-ax} \sin(bx + c) + be^{-ax} \cos(bx + c)$$

Test yourself

Differentiate the following functions with respect to x.

1 e^{5x}

2 $e^{\frac{1}{2}x}$

3 $e^{\sqrt{x}}$

4 e^{-2x}

5 $e^{-\frac{5}{2}x}$

6 $e^{(5-2x)}$

7 $e^{(ax+b)}$

8 e^{x^2}

9 e^{-px}

10 $e^{\frac{x}{a}}$

11 $\dfrac{e^x + e^{-x}}{2}$

12 $\dfrac{e^x - e^{-x}}{2}$

13 xe^x

14 xe^{-x}

15 $x^2 e^{-x}$

16 $(x + 4)e^x$

17 $e^x \sin x$

18 $10e^x$

19 2^x

20 10^{2x}

21 $x^n a^x$

22 $e^{\sin x}$

23 $(a + b)^x$

24 a^{2x+1}

25 a^{bx^2}

26 $e^{\cos x}$

27 $e^{\tan x}$

28 $\ln \frac{x}{a}$

29 $\ln(ax^2 + bx + c)$

30 $\ln x^2$

31 $\ln(x^3 + 3)$

32 $x \ln x$

33 $\ln(px + q)$

34 $\ln \sin x$

35 $\ln \cos x$

36 $\ln \left(\dfrac{a + x}{a - x} \right)$

37 $\ln(e^x + e^{-x})$ **38** $\ln\left(x + \sqrt{x^2 + 1}\right)$ **39** $\sqrt{x} - \ln\left(1 + \sqrt{x}\right)$

40 $\ln \tan \frac{1}{2}x$ **41** $\ln\left(\sqrt{x^2 + 1}\right)$ **42** $\dfrac{e^x}{\sqrt{x}}$

43 $x^2 e^{4x}$ **44** $ae^{-kx} \sin kx$ **45** $\ln\sqrt{\sin x}$

46 x^2 **47** $\ln\left(\sqrt{x-1} + \sqrt{x+1}\right)$ **48** $\ln\dfrac{e^x}{1 + e^x}$

49 $\ln\dfrac{x}{a - \sqrt{a^2 - x^2}}$ **50** $\sin^{-1}\dfrac{e^x - e^{-x}}{e^x + e^{-x}}$

Key ideas

▶ The *exponential function* describes a situation where a quantity grows or decays at a rate proportional to its current value.

▶ Algebraically, the simplest form of the exponential function is $y = e^x$.

▶ A useful model for thinking about the exponential function is the formula for compound interest.

▶ A key property of the function e^x is that its derivative is also e^x. The derivative of the function e^{ax} is $a \times e^{ax}$.

▶ The logarithm function is the inverse of the exponential function. Thus, if $y = e^x$ then $x = log_e y$. Note that log_e means log 'to the base e' and is usually referred to as a natural logarithm (often written ln).

10

Hyperbolic functions

In this chapter you will learn:

▶ *the definitions of the hyperbolic functions,* sinh *x and* cosh *x in terms of* e^x

▶ *how to differentiate* sinh *x and* cosh *x*

▶ *that the properties of* sinh *x and* cosh *x have similarities with those of* sin *x and* cos *x.*

10.1 Definitions of hyperbolic functions

Figure 9.1 showed the graphs of the exponential functions $y = e^x$ and $y = e^{-x}$. These two curves are reproduced in Figure 10.1, together with two other curves marked **A** and **B**.

▶ For curve **A**, the y-value of any point on it is one half of the sum of the corresponding y-values of e^x and e^{-x}. For example, at the point P, its ordinate PQ is half the sum of LQ and MQ.

Therefore for every point on curve **A**:

$$y = \frac{e^x + e^{-x}}{2}$$

▶ On curve **B**, the y-value of any point is one half of the difference of the y-values of the other two curves.

Thus $RQ = \frac{1}{2}(LQ - MQ)$, that is, for every point,

$$y = \frac{e^x - e^{-x}}{2}$$

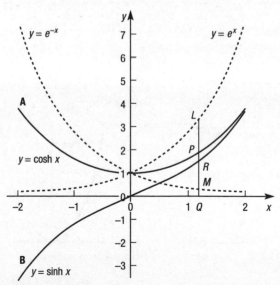

Figure 10.1 Graphs of y sinh x and $y = \cosh x$

The two graphs therefore represent two functions of x, and their equations are given by

$$y = \frac{1}{2}(e^x + e^{-x}) \text{ and } y = \frac{1}{2}(e^x - e^{-x})$$

These two functions have properties which in many respects are very similar to those of $y = \cos x$ and $y = \sin x$. It can be shown that they bear a similar relation to the hyperbola, as the trigonometrical or circular functions do to the circle. Hence the function $y = \frac{1}{2}(e^x + e^{-x})$ is called the **hyperbolic cosine**, and $y = \frac{1}{2}(e^x - e^{-x})$ is called the **hyperbolic sine**.

These names are abbreviated to $\cosh x$ and $\sinh x$, the added 'h' indicating the hyperbolic cosine and hyperbolic sine. The names are usually pronounced 'cosh' and 'shine' respectively.

Definition: $\cosh x = \frac{1}{2}(e^x + e^{-x})$ and $\sinh x = \frac{1}{2}(e^x - e^{-x})$

Nugget – Building up sinh and cosh yourself

As you will already be aware, these ideas can seem very abstract if all you do is read about the hyperbolic functions. Much better for you is to be an active learner and try to create these functions for yourself. You can do this using a computing application like Geogebra or a graphing calculator.

On the screen illustrated below, the functions $Y_1 = e^x$ and $Y_2 = e^{-x}$ have been entered on the $Y=$ screens of a graphing calculator. The graphs of these two functions are shown alongside.

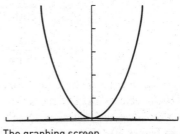

```
Plot 1    Plot 2    Plot 3
\Y₁ = e^(x)
\Y₂ = e^(-x)
\Y₃ =
\Y₄ =
\Y₅ =
\Y₆ =
\Y₇ =
```

Figure 10.2 The $Y=$ screen   The graphing screen

Now a third function, Y_3 is created which is half the sum of the first two functions.

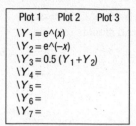
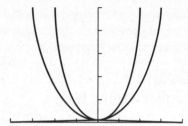

Plot 1	Plot 2	Plot 3

$\backslash Y_1 = e^\wedge(x)$
$\backslash Y_2 = e^\wedge(-x)$
$\backslash Y_3 = 0.5\,(Y_1 + Y_2)$
$\backslash Y_4 =$
$\backslash Y_5 =$
$\backslash Y_6 =$
$\backslash Y_7 =$

Figure 10.3

This is the *cosh* function, $y = \cosh x$

You can continue in this manner, entering a fourth function. Y_4 can be created $Y_4 = 0.5(Y_1 - Y_2)$, i.e. which is half the *difference* of the first two functions, and this will be the *sinh* function, $y = \sinh x$.

From these definitions:

$$\cosh x + \sinh x = e^x \text{ and } \cosh x - \sinh x = e^{-x}$$

There are four other hyperbolic functions corresponding to the other circular functions, namely:

$$\tanh x = \frac{\sinh x}{\cosh x} = \frac{e^x - e^{-x}}{e^x + e^{-x}} = \frac{e^{2x} - 1}{e^{2x} + 1}$$

$$\operatorname{cosech} x = \frac{1}{\sinh x} = \frac{2}{e^x - e^{-x}}$$

$$\operatorname{sech} x = \frac{1}{\cosh x} = \frac{2}{e^x + e^{-x}}$$

$$\coth x = \frac{1}{\tan x} = \frac{e^x + e^{-x}}{e^x - e^{-x}}$$

You can express these functions in exponential form by using their reciprocals. The names of these functions are pronounced 'than', 'coshec', 'shec' and 'coth'. The graph of cosh x, marked **A** in Figure 10.1, is especially important. It is called the **catenary**, and is the curve formed by a uniform flexible chain which hangs freely with its ends fixed.

10.2 Formulae connected with hyperbolic functions

There is a close correspondence between formulae expressing relations between hyperbolic functions, and similar relations between circular functions.

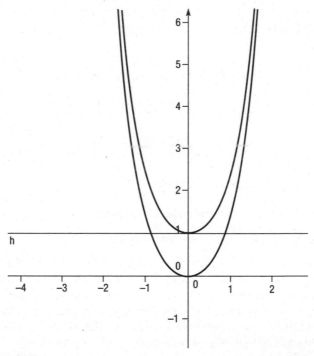

Consider the two following examples.

Example 10.2.1

$$\cosh^2 x - \sinh^2 x = \left(\frac{e^x + e^{-x}}{2}\right)^2 - \left(\frac{e^x - e^{-x}}{2}\right)^2$$

$$= \frac{1}{4}\left\{(e^{2x} + e^{-2x} + 2) - (e^{2x} + e^{-2x} - 2)\right\}$$

$$= 1$$

Therefore $\cosh^2 x - \sinh^2 x = 1$

Compare this result with the trigonometrical result

$$\cos^2 x + \sin^2 x = 1$$

Example 10.2.2

$$\cosh^2 x + \sinh^2 x = \left(\frac{e^x + e^{-x}}{2}\right)^2 + \left(\frac{e^x - e^{-x}}{2}\right)^2$$

$$= \left(\frac{e^{2x} + e^{-2x} + 2}{4}\right) + \left(\frac{e^{2x} + e^{-2x} - 2}{4}\right)$$

$$= \left(\frac{2e^{2x} + 2e^{-2x}}{4}\right) = \frac{e^{2x} - e^{2x}}{2}$$

$$= \cosh 2x$$

Therefore $\cosh^2 x + \sinh^2 x = \cosh 2x$

This is analogous to $\cos^2 x - \sin^2 x = \cos 2x$

Similarly, every formula for circular functions has its counterpart in hyperbolic functions. Notice that in the above two cases there is a difference in the signs used, and that this difference applies only to sin x. This has led to the formulation of Osborne's rule, by which you can write down formulae for hyperbolic functions immediately from the corresponding formulae for circular functions.

OSBORNE'S RULE

In any formula connecting circular functions of general angles, the corresponding formula connecting hyperbolic functions can be obtained by replacing each circular function by the corresponding hyperbolic function, if the sign of every product or implied product of two sines is changed.

For example $\sec^2 x = 1 + \tan^2 x$ becomes $\operatorname{sech}^2 x = 1 - \tanh^2 x$

since $$\tanh^2 x = \frac{\sinh x \times \sinh x}{\cosh x \times \cosh x}$$

10.3 Summary

The more important of these corresponding formulae are summarized for convenience.

Hyperbolic functions	Circular functions
$\cosh^2 x - \sinh^2 x = 1$	$\cos^2 x + \sin^2 x = 1$
$\sinh 2x = 2 \sinh x \cosh x$	$\sin 2x = 2 \sin x \cos x$
$\cosh 2x = \cosh^2 x + \sinh^2 x$	$\cos 2x = \cos^2 x - \sin^2 x$
$\operatorname{sech}^2 x = 1 - \tanh^2 x$	$\sec^2 x = 1 + \tan^2 x$
$\operatorname{cosech}^2 x = \coth^2 x - 1$	$\operatorname{cosec}^2 x = \cot^2 x + 1$
$\sinh(x + y) = \sinh x \cosh y + \cosh x \sinh y$	$\sin(x + y) = \sin x \cos y + \cos x \sin y$
$\sinh(x - y) = \sinh x \cosh y - \cosh x \sinh y$	$\sin(x - y) = \sin x \cos y - \cos x \sin y$
$\cosh(x + y) = \cosh x \cosh y + \sinh x \sinh y$	$\cos(x + y) = \cos x \cos y - \sin x \sin y$
$\cosh(x - y) = \cosh x \cosh y - \sinh x \sinh y$	$\cos(x - y) = \cos x \cos y + \sin x \sin y$

The following striking connections between the two sets of functions are given for your information. For a fuller treatment consult a book on advanced trigonometry.

$$\cosh x = \frac{1}{2}(e^x + e^{-x}) \qquad \cos x = \frac{1}{2}(e^{ix} + e^{-ix})$$

$$\sinh x = \frac{1}{2}(e^x - e^{-x}) \qquad \sin x = \frac{1}{2i}(e^{ix} - e^{-ix})$$

$$\sinh x = \frac{1}{i}\sin ix$$

$$\cosh x = \cos ix$$

Nugget – Complex numbers

In case your knowledge of complex numbers is rusty, here are some pointers. Complex numbers were first thought up by the Italian mathematician Gerolamo Cardano when attempting to find solutions to cubic equations. At the time he named them *fictitious* numbers, because they really only existed in his imagination.

A complex number would result from solving a quadratic equation of the form $y = ax^2 + bx + c$ under circumstances where $b^2 < 4ac$, because the *discriminant* of the quadratic formula, $b^2 - 4ac$, would thereby be negative and this would result in having to find the square root of a negative number.

So, a complex number involves taking the square root of a negative number – for example $\sqrt{(-4)}$. Note that $\sqrt{(-4)}$ can be rewritten as $\sqrt{(4 \times -1)}$, which simplifies to $2\sqrt{(-1)}$. By this sort of simplification, all such numbers can be reduced to 'some real number' multiplied by $\sqrt{(-1)}$. For convenience, $\sqrt{(-1)}$ is given the letter i, so $\sqrt{(-4)}$ can be expressed as $2i$.

10.4 Derivatives of the hyperbolic functions

DERIVATIVE OF sinh x

Let $\quad y = \sinh x = \dfrac{e^x - e^{-x}}{2}$ $\qquad$ Then $\quad \dfrac{dy}{dx} = \dfrac{e^x + e^{-x}}{2} = \cosh x$

DERIVATIVE OF cosh x

Let $\quad y = \cosh x = \dfrac{e^x + e^{-x}}{2}$ $\qquad$ Then $\quad \dfrac{dy}{dx} = \dfrac{e^x - e^{-x}}{2} = \sinh x$

DERIVATIVE OF tanh x

The derivative may be found from the definition in terms of the exponential function, but it is easier to use the results proved above.

Let $\qquad\qquad\qquad y = \tanh x = \dfrac{\sinh x}{\cosh x}$

Then

$$\frac{dy}{dx} = \frac{\cosh x \times \cosh x - \sinh x \times \sinh x}{\cosh^2 x}$$

$$= \frac{\cosh^2 x - \sinh^2 x}{\cosh^2 x}$$

$$= \frac{1}{\cosh^2 x}$$

$$= \operatorname{sech}^2 x$$

Similarly, you can show that:

If $y = \operatorname{cosech} x$ then $\dfrac{dy}{dx} = \operatorname{cosech} x \coth x$

If $y = \operatorname{sech} x$ then $\dfrac{dy}{dx} = -\operatorname{sech} x \tanh x$

If $y = \coth x$ then $\dfrac{dy}{dx} = -\operatorname{cosech}^2 x$

Compare these results with the derivatives of the corresponding circular functions.

10.5 Graphs of the hyperbolic functions

Look again at the graphs of $\cosh x$ and $\sinh x$ in Figure 10.1, now that you know the derivatives of these functions.

GRAPH OF $y = \cosh x$, $\dfrac{dy}{dx} = \sinh x$, $\dfrac{d^2y}{dx^2} = \cosh x$

The derivative $\frac{dy}{dx} = \sinh x$ vanishes only when $x = 0$. There is therefore a turning point on curve **A**. Also since $\sinh x$ is negative before $x = 0$ and positive after $x = 0$ and $\frac{d^2y}{dx^2} = \cosh x$ is positive, the point is a local minimum. There is no other turning point and no point of inflexion.

GRAPH OF $y = \sinh x$, $\dfrac{dy}{dx} = \cosh x$, $\dfrac{d^2y}{dx^2} = \sinh x$

The derivative $\frac{dy}{dx} = \cosh x$ is always positive and does not vanish. Consequently $\sinh x$ has no turning point. When $x = 0$,

$\frac{d^2y}{dx^2} = 0$ and is negative before and positive after $x = 0$. Therefore there is a point of inflexion when $x = 0$. Since $\frac{dy}{dx} = \cosh x = 1$ when $x = 0$, the gradient is 1 when $x = 0$.

GRAPH OF $y = \tanh x,\ \dfrac{dy}{dx} = \text{sech}^2 x$

Since $\text{sech}^2 x$ is always positive, $\tanh x$ is increasing for all values of x. Also since $\sinh x$ and $\cosh x$ are always continuous and $\cosh x$ never vanishes, $\tanh x$ must be a continuous function.

In Section 10.1, $\tanh x$ was written in the form:

$$\tanh x = \frac{e^{2x} - 1}{e^{2x} + 1} = 1 - \frac{2}{e^{2x} + 1}$$

From this form you can see that while x increases from $-\infty$ to 0, e^{2x} increases from 0 to 1.

Therefore $1 - \dfrac{2}{e^{2x} + 1}$, that is, $\tanh x$, increases from -1 to 0.

Similarly, as x increases from 0 to ∞, $\tanh x$ increases from 0 to 1. The graph therefore has the lines $y = 1$ and $y = -1$ as its asymptotes. The graph of $y = \tanh x$ is shown in Figure 10.5.

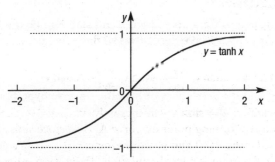

Figure 10.5 Graph of $y = \tanh x$

10.6 Differentiating the inverse hyperbolic functions

Inverse hyperbolic functions correspond to inverse circular functions, and you find their derivatives by similar methods.

DERIVATIVE OF sinh^{-1} x

Let $\quad y = \sinh^{-1} x \quad$ Then $\quad x = \sinh y$

$$\frac{dy}{dx} = \frac{1}{\cosh y} = \frac{1}{\sqrt{1 + \sinh^2 y}}$$

$$= \frac{1}{\sqrt{1 + x^2}}$$

You take the positive value of the square root because $\cosh y$ is always positive.

DERIVATIVE OF cosh^{-1} x

Using the same method as above you get

$$y = \cosh^{-1} x$$

$$\frac{dy}{dx} = \frac{1}{\sqrt{x^2 - 1}}$$

DERIVATIVE OF tanh^{-1} x

Let $\quad y = \tanh^{-1} x \quad$ Then $\quad x = \tanh y$

$$\frac{dx}{dy} = \operatorname{sech}^2 y$$

$$\frac{dy}{dx} = \frac{1}{\operatorname{sech}^2 y} = \frac{1}{1 - \tanh^2 y} \qquad \text{(Section 10.3)}$$

$$= \frac{1}{1 - x^2}$$

You can find the derivatives of the other inverse hyperbolic functions by similar methods.

They are:

$$y = \operatorname{sech}^{-1} x \qquad \frac{dy}{dx} = -\frac{1}{x\sqrt{1 - x^2}}$$

$$y = \operatorname{cosech}^{-1} x \qquad \frac{dy}{dx} = -\frac{1}{x\sqrt{1 + x^2}}$$

$$y = \coth^{-1} x \qquad \frac{dy}{dx} = -\frac{1}{x^2 - 1}$$

You will find the following forms important later in integration.

If
$$y = \sinh^{-1} \frac{x}{a}$$

$$\frac{dy}{dx} = \frac{1}{\sqrt{1 + \left(\frac{x}{a}\right)^2}} \times \frac{1}{a} = \frac{1}{\sqrt{a^2 + x^2}}$$

Similarly, if
$$y = \cosh^{-1} \frac{x}{a}$$

$$\frac{dy}{dx} = \frac{1}{\sqrt{x^2 - a^2}}$$

10.7 Logarithm equivalents of the inverse hyperbolic functions

Each of the inverse hyperbolic functions has an equivalent form using natural logarithms.

THE RELATION $\sinh^{-1} x = \ln(x + \sqrt{1 + x^2})$

Let
$$y = \sinh^{-1} x$$

Then
$$x = \sinh y$$

Now
$$\cosh y + \sinh y = e^y \qquad \text{(Section 10.1)}$$

and
$$\cosh^2 y - \sinh^2 y = 1 \qquad \text{(Section 10.2)}$$

It follows that
$$e^y = x + \sqrt{1 + x^2}$$

where the positive sign is taken because $\cosh y$ is always positive.

Taking logarithms gives
$$y = \ln\left(x + \sqrt{1 + x^2}\right)$$

or
$$\sinh^{-1} x = \ln\left(x + \sqrt{1 + x^2}\right)$$

THE RELATION $\cosh^{-1} x = \ln \left(x + \sqrt{x^2 - 1} \right)$

Let $\qquad\qquad\qquad\qquad y = \cosh^{-1} x$

Then $\qquad\qquad\qquad\qquad x = \cosh y$

But $\qquad\qquad\qquad\qquad \sinh^2 y = \cosh^2 y - 1 \qquad$ (Section 10.2)

$\qquad\qquad\qquad\qquad\qquad\quad = x^2 - 1$

As above $\qquad\qquad\qquad e^y = \cosh y + \sinh y$

so $\qquad\qquad\qquad\qquad e^y = x \pm \sqrt{x^2 - 1}$

In this both signs apply, since $\sinh y$ can be positive or negative.

Therefore $\qquad\qquad\qquad y = \ln \left(x \pm \sqrt{x^2 - 1} \right)$

But $\cosh^{-1} x$ is a function, and can only take one value for each value of x. It is usual to take the positive value.

Thus $\qquad\qquad\qquad \cosh^{-1} x = \ln \left(x + \sqrt{x^2 - 1} \right)$

If you need the negative value of y such that $\cosh y = x$ you

should take $y = -\ln \left(x + \sqrt{x^2 - 1} \right)$

Notice that

$$-\ln \left(x + \sqrt{x^2 - 1} \right) = \ln \left(x + \sqrt{x^2 + 1} \right)^{-1}$$

$$= \ln \left(\frac{1}{x + \sqrt{x^2 - 1}} \right)$$

$$= \ln \left(\frac{1}{x + \sqrt{x^2 - 1}} \times \frac{x - \sqrt{x^2 - 1}}{x - \sqrt{x^2 - 1}} \right)$$

$$= \ln \left(x - \sqrt{x^2 - 1} \right)$$

Note also that x must be greater than 1.

THE RELATION $\tanh^{-1} x = \frac{1}{2} \ln \dfrac{1+x}{1-x}$

Let $\quad y = \tanh^{-1} x \qquad\qquad$ so $\quad x = \tanh y$

and x lies between -1 and 1 (see Section 10.5).

Now $x = \dfrac{e^{2y} - 1}{e^{2y} + 1}$ Therefore $x(e^{2y} + 1) = e^{2y} - 1$ so $e^{2y} = \dfrac{1 + x}{1 - x}$

(Section 10.5)

Therefore $\quad 2y = \ln\dfrac{1 + x}{1 - x}$ and $\quad y = \dfrac{1}{2}\ln\dfrac{1 + x}{1 - x}$

Therefore $\quad \tanh^{-1}x = \dfrac{1}{2}\ln\dfrac{1 + x}{1 - x}$

10.8 Summary of inverse functions

Function	Derivative
$\sinh^{-1}x$	$\dfrac{1}{\sqrt{1 + x^2}}$
$\cosh^{-1}x$	$\dfrac{1}{\sqrt{x^2 - 1}}$
$\tanh^{-1}x$	$\dfrac{1}{1 - x^2}$
$\operatorname{cosech}^{-1}x$	$\dfrac{-1}{x\sqrt{x^2 + 1}}$
$\operatorname{sech}^{-1}x$	$\dfrac{-1}{x\sqrt{1 - x^2}}$
$\coth^{-1}x$	$\dfrac{-1}{x^2 - 1}$

The following additional forms are important.

Function	Derivative
$y = \sinh^{-1}\dfrac{x}{a}$	$\dfrac{1}{\sqrt{x^2 + a^2}}$
$y = \cosh^{-1}\dfrac{x}{a}$	$\dfrac{1}{\sqrt{x^2 - a^2}}$
$y = \tanh^{-1}\dfrac{x}{a}$	$\dfrac{a}{a^2 - x^2}$

LOGARITHM EQUIVALENTS

$$\sinh^{-1}x = \ln\left(x + \sqrt{1+x^2}\right)$$

$$\cosh^{-1}x = \ln\left(x + \sqrt{x^2-1}\right)$$

$$\tanh^{-1}x = \frac{1}{2}\ln\frac{1+x}{1-x}$$

$$\sinh^{-1}\frac{x}{a} = \ln\left(\frac{x + \sqrt{a^2+x^2}}{a}\right)$$

$$\cosh^{-1}\frac{x}{a} = \ln\left(\frac{x + \sqrt{x^2-a^2}}{a}\right)$$

$$\tanh^{-1}\frac{x}{a} = \frac{1}{2}\ln\frac{a+x}{a-x}$$

Test yourself

Differentiate the following functions.

1 $\sinh\frac{1}{2}x$	**2** $\sinh 2x$	**3** $\cosh\frac{1}{3}x$
4 $\tanh ax$	**5** $\tanh\frac{1}{4}x$	**6** $\sinh ax + \cosh ax$
7 $\sinh\frac{1}{x}$	**8** $\sinh^2 x$	**9** $\cosh^3 x$
10 $\sinh(ax + b)$	**11** $\cosh 2x^2$	**12** $\sinh^n ax$
13 $\sinh x \cosh x$	**14** $\sinh^2 x + \cosh^2 x$	**15** $\tanh^2 x$
16 $\ln \tanh x$	**17** $x \sinh x - \cosh$	**18** $\ln \cosh x$
19 $x^3 \sinh 3x$	**20** $\ln(\sinh x + \cosh x)$	**21** $e^{\sinh x}$
22 $\sqrt{\sinh x}$	**23** $\ln\frac{1 + \tanh x}{1 - \tanh x}$	**24** $e^{\tanh x}$
25 $\sinh^{-1}\frac{1}{2}x$	**26** $\cosh^{-1}\frac{1}{5}x$	**27** $\sinh^{-1}\frac{1-x}{1+x}$
28 $\sinh^{-1}\tan x$	**29** $\tan^{-1}\sinh x$	**30** $\tanh^{-1}\sin x$
31 $\sin^{-1}\tanh x$	**32** $\cosh^{-1}\sec x$	**33** $\tanh^{-1}\frac{2x}{1+x^2}$

34 $\tanh^{-1}\frac{1}{1+x}$

35 $\sinh^{-1}\left(2x\sqrt{1+x^2}\right)$

36 $\cosh^{-1}(4x+1)$

37 $\sinh^{-1}\tan\frac{1}{2}x$

38 $\sin^{-1}\tanh\frac{1}{2}x$

Write the logarithmic equivalents of the following.

39 $\sinh^{-1}\frac{1}{2}x$

40 $\cosh^{-1}\frac{1}{3}x$

41 $\cosh^{-1}\frac{3}{2}x$

42 $\tanh^{-1}\frac{1}{4}x$

Key ideas

▶ *Hyperbolic functions* have certain properties that are similar to the trigonometric functions, sin, cos and tan.

▶ The three basic hyperbolic functions are the hyperbolic sine 'sinh', the hyperbolic cosine 'cosh' and the hyperbolic tangent 'tanh'.

▶ The hyperbolic functions are represented algebraically by equations that derive from the exponential function. For example, hyperbolic sine x, $\sinh x = \frac{1}{2}(e^x - e^{-x})$.

▶ *Osborne's rule* (Section 10.2) shows a neat connection between any formula connecting trigonometric functions of general angles and the corresponding formula connecting hyperbolic functions.

▶ The derivatives of the six hyperbolic functions are given in Section 10.4.

▶ The derivatives of the six inverse hyperbolic functions are summarized in a table in Section 10.8.

11

Integration;
standard integrals

In this chapter you will learn:

▶ *that the reverse of differentiation is called integration*

▶ *the notation for integration*

▶ *how to integrate simple expressions by inspection.*

11.1 Meaning of integration

The integral calculus is concerned with the operation of **integration**, which, in one of its aspects, is the **reverse of differentiation**.

From this point of view, the problem to be solved in integration is: *What is the function which, on being differentiated, produces a given function?* For example, what is the function which, being differentiated, produces $\cos x$? In this case you know (from the work in the previous chapters on differentiation), that $\sin x$ is the function required. Therefore $\sin x$ is the **integral** of $\cos x$.

Generally if $f'(x)$ represents the derivative of $f(x)$, then the problem of integration is, given $f'(x)$, find $f(x)$, or given $\frac{dy}{dx}$, find y.

But the process of finding the integral is seldom as simple as in the example above. A converse operation is usually more difficult than the direct one, and integration is no exception. A sound knowledge of differentiation will help in many cases (as in that above) but even when the type of function is known, there may arise minor complications of signs and constants. For example, if you want the integral of $\sin x$ you know that $\cos x$, when differentiated, produces $-\sin x$. You therefore conclude that the function which produces $\sin x$ on differentiation must be $-\cos x$. Thus the integral of $\sin x$ is $-\cos x$.

Again, suppose you want the integral of x. You know that the function which produces x on differentiation must be of the form x^2. But $\frac{d}{dx}(x^2) = 2x$. If the result of differentiation is to be x, the integral must be a multiple of x so that the multiple cancels with the 2 in $2x$. Clearly this constant must be $\frac{1}{2}$. Hence the integral must be $\frac{1}{2}x^2$.

These two examples may help you to realize some of the difficulties which face you in the integral calculus. In the differential calculus (with a knowledge of the rules from the previous chapters), it is possible to differentiate not only all the ordinary types of functions, but also complicated expressions formed by products, powers, quotients, logarithms, and so on, of these functions. But simplifications, cancellings and other operations occur before the final form of the derivative is reached. When reversing the process, as in integration, you want to know the original function. It is usually impossible to reverse all of these changes, and in many cases you cannot carry out the integration.

It is not possible to formulate a set of rules by which any function may be integrated. However, methods have been devised for integrating certain types of functions, and these will be stated in succeeding chapters. Knowing these methods and practising them, if you possess a good grasp of differentiation, will enable you to integrate most of the commonly occurring functions.

In general, these methods consist of transposing and manipulating the functions, so that they assume known forms of standard functions of which the integrals are known. The final solution becomes a matter of recognition and inspection.

Integration has one advantage – you can always check the result. If you differentiate the function you obtain by integration, you should get the original function. You should not omit this check.

Nugget – The first thousand are the hardest...

As you will discover in this chapter, finding the integral of certain functions can be more of an art than a science. Being good at integration requires you to be good at spotting certain patterns and this is achieved by making use of previous knowledge and experience. These awarenesses will gradually emerge by spending a lot of time working up your skills at differentiation as well as integration.

In short, there are three main ways of becoming good at integration and these are:

▶ practice

▶ practice, and

▶ practice.

As one of my teachers was fond of observing to his students, 'Don't worry; by the time you've done your first thousand integrals, you'll find that you are starting to get the hang of it!'

11.2 The constant of integration

When a function containing a constant term is differentiated, the constant term disappears, since its derivative is zero. When the process is reversed in integration, the constant cannot be determined without further information.

For example, when $y = x^2 + 3$

you know that $\dfrac{dy}{dx} = 2x$

If you reverse the process and integrate $2x$ as it stands, the result is x^2. Consequently to get a complete integral you must add an unknown constant.

In the above example, let C denote the constant. Then you can say that the integral of $2x$ is $x^2 + C$, where C is an undetermined constant, often called an **arbitrary constant**. Consequently, the integral is called an **indefinite integral**.

You can illustrate this graphically as follows. In Figure 11.1 the graphs of $y = x^2$, $y = x^2 + 2$ and $y = x^2 - 3$ are all included in the general form $y = x^2 + C$. They are called **integral curves**, since they represent the curves of the integral $x^2 + C$, when the values 0, 2 and -3 are assigned to C. There are infinitely many such curves.

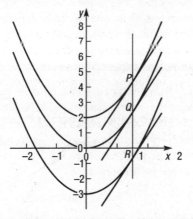

Figure 11.1

Let P, Q and R be points on these curves where they are cut by the line $x = 1.5$, which is parallel to the y-axis.

At all three points the derivative is $2x$ and the three points also all have the same x-coordinate, 1.5. Therefore they all have the same gradient, 3.

The integral $y = x^2 + C$ therefore represents a set of curves all having the same gradients at points with the same value of x.

You can find the equation of any particular curve in the set if you know a pair of corresponding values of x and y. These enable you to find C. For example, if a curve passes through the point $(3, 6)$, these values of x and y can be substituted in the equation.

Thus on substitution in: $\qquad\qquad\qquad y = x^2 + C$

you find that: $\qquad\qquad\qquad\qquad 6 = 3^2 + C$

giving: $\qquad\qquad\qquad\qquad\qquad\quad C = -3$

Thus $y = x^2 - 3$ is the equation of this particular curve in the set.

11.3 The symbol for integration

The operation of integration requires a symbol to indicate it. The symbol chosen is:

$$\int$$

which is the old-fashioned, elongated 's', selected because it is the first letter of the word 'sum', which, as you will see in Section 15.2, is another aspect of integration.

The differential dx is written next to the function to be integrated in order to indicate the independent variable used for the original differentiation, and the variable which is to be used for the integration. Thus, $\int f(x)dx$ means that $f(x)$ is to be integrated with respect to x. The integral of $\cos x$ (considered in Section 11.1) would be written as

$$\int \cos x \, dx = \sin x + C$$

It is important to remember that the variables in the function to be integrated and in the differential must be the same. Thus

$$\int \cos y \, dx$$

could not be integrated as it stands. You would first have to express $\cos y$ as a function of x.

Note that you can use any other letter besides x to indicate the independent variable. Thus $\int t\, dt$ indicates that t is the independent variable, and that you need to integrate t with respect to it.

11.4 Integrating a constant factor

It was shown in Example 5.6.5 that when a function contains a constant number as a factor, this number will also be a factor of the derivative of the function. Thus if

$$y = ax^n$$

then

$$\frac{dy}{dx} = a(nx^{n-1})$$

It will be obvious from Example 5.6.5 that when the operation is reversed, and you integrate a function containing a constant factor, this factor must also be a factor of the final integral.

When finding an integral it is better to transfer such a factor to the left side of the integration sign before proceeding with the integration of the function.

Thus

$$\int 5x\, dx = 5 \int x\, dx$$
$$= 5(\tfrac{1}{2}x^2) + C$$
$$= \tfrac{5}{2}x^2 + C$$

Generally

$$\int af'(x)\, dx = a \int f'(x)\, dx$$

Note that you cannot transfer a factor which involves the variable to the other side of the integration sign in the same way as a constant.

11.5 Integrating x^n

You can find simple examples of this integral by inspection, namely

$$\int x\, dx = \tfrac{1}{2} x^2 + C$$

$$\int x^2\, dx = \tfrac{1}{3} x^3 + C$$

$$\int x^3\, dx = \tfrac{1}{4} x^4 + C$$

$$\int x^4\, dx = \tfrac{1}{5} x^5 + C$$

From these examples you can deduce that

$$\int x^n\, dx = \frac{1}{n+1} x^{n+1} + C$$

Also, in accordance with the rule of Section 11.4,

$$\int a x^n\, dx = a \int x^n\, dx$$

$$= \frac{a}{n+1} x^{n+1} + C$$

From the function of a function rule, you can also deduce that

$$\int (ax + b)^n\, dx = \frac{1}{a(n+1)} (ax + b)^{n+1} + C$$

If you have difficulty in understanding this result, you can check it by differentiating the integral obtained.

The rule for the differentiation of x^n holds for all values of n. The formula for integrating x^n holds for all n except $n = -1$. The case $n = -1$ is considered in Section 11.7. Note that:

$$\int dx = x + C$$

Example 11.5.1

$$\int 3x^7\, dx = 3\int x^7\, dx = 3 \times \frac{x^8}{8} + C$$
$$= \frac{8}{3} x^8 + C$$

Example 11.5.2

$$\int 4\sqrt{x}\, dx = 4\int x^{\frac{1}{2}}\, dx = 4 \times \frac{x^{\frac{1}{2}+1}}{\frac{1}{2}+1} + C = 4 \times \frac{2}{3} x^{\frac{3}{2}} + C$$
$$= \frac{8}{3} x^{\frac{3}{2}} + C$$

Example 11.5.3

$$\int \frac{dx}{\sqrt{x}} = \int x^{-\frac{1}{2}}\, dx = \frac{x^{-\frac{1}{2}+1}}{-\frac{1}{2}+1} + C = 2x^{\frac{1}{2}} + C$$
$$= 2\sqrt{x} + C$$

Example 11.5.4

$$\int \frac{dx}{\sqrt{x+b}} = \int (x+b)^{-\frac{1}{2}}\, dx = \frac{(x+b)^{-\frac{1}{2}+1}}{-\frac{1}{2}+1} + C$$

$$= 2(x+b)^{\frac{1}{2}} + C = 2\sqrt{x+b} + C$$

Example 11.5.5

$$\int \frac{dx}{\sqrt{ax+b}} = \int (ax+b)^{-\frac{1}{2}}\, dx = \frac{1}{a} \times \frac{(ax+b)^{-\frac{1}{2}+1}}{-\frac{1}{2}+1} + C$$

$$= \frac{2}{a}(ax+b)^{\frac{1}{2}} + C = \frac{2}{a}\sqrt{ax+b} + C$$

11.6 Integrating a sum

By considering the rule for differentiating the sum of a number of functions, the same rule must hold for integration, that is, the integral of a sum of functions is equal to the sum of the integrals of these functions.

Example 11.6.1

Find $\int (x^3 - 5x^2 + 7x - 11) dx$

$$\int (x^3 - 5x^2 + 7x - 11) dx = \int x^3 dx - 5 \int x^2 dx + 7 \int x dx - 11 \int dx$$

$$= \frac{x^4}{4} - \frac{5x^3}{3} + \frac{7x^2}{2} - 11x + C$$

Note: The constants that would arise from the integration of the separate terms can all be included in **one** constant, since this constant is arbitrary and undetermined. In practice most people would not put in the first step, but carry it out mentally.

Example 11.6.2

Find $\int \left(\sqrt[3]{x} - \frac{1}{\sqrt[3]{x}} \right) dx$

$$\int \left(\sqrt[3]{x} - \frac{1}{\sqrt[3]{x}} \right) dx = \int x^{\frac{1}{3}} dx - \int x^{-\frac{1}{3}} dx$$

$$= \frac{1}{\frac{1}{3} + 1} x^{\frac{4}{3}} - \frac{1}{-\frac{1}{3} + 1} x^{\frac{2}{3}} + C$$

$$= \frac{3}{4} x^{\frac{4}{3}} - \frac{3}{2} x^{\frac{2}{3}} + C$$

11.7 Integrating $\frac{1}{x}$

If you apply the rule for integrating x^n to $\frac{1}{x}$, that is, to x^{-1}, you find

$$\int \frac{1}{x}\,dx = \int x^{-1}\,dx = \frac{x^{-1+1}}{-1+1} + C$$

which involves a division by zero. This shows that the rule

$$\int x^n dx = \frac{1}{n+1} x^{n+1} + C$$

does not work in the case $n = -1$.

You know, however, that the rule for differentiating the logarithm function is:

$$\frac{d}{dx}(\ln x) = \frac{1}{x} \qquad\qquad \text{for } x > 0$$

Hence you can conclude that:

$$\int \frac{1}{x}\,dx = \ln x + C \qquad\qquad \text{for } x > 0$$

Note that this rule applies only when $x > 0$

However, when you differentiate $\ln(-x)$, which is defined only for $x < 0$, you find that:

$$\frac{d}{dx}(\ln(-x)) = \frac{-1}{-x} = \frac{1}{x} \qquad\qquad \text{for } x < 0$$

so $\qquad \int \frac{1}{x}\,dx = \ln(-x) + C \qquad\qquad \text{for } x < 0$

You can combine these two formulae for integrating $\frac{1}{x}$ by writing:

$$\int \frac{1}{x}\,dx = \ln|x| + C$$

where $|x|$ is the **modulus** of x, and is defined by:

$$|x| = x \text{ if } x \geq 0$$

$$|x| = -x \text{ if } x < 0$$

11.8 A useful rule for integration

By combining the rule for differentiating a function of a function with the last result you find that if:

$$y = \ln|f(x)|$$

then

$$\frac{dy}{dx} = \frac{1}{f(x)} \times f'(x) = \frac{f'(x)}{f(x)}$$

Consequently

$$\int \frac{f'(x)}{f(x)} dx = \ln|f(x)| + C$$

Hence, when integrating a rational (or fractional) function in which, (after a suitable adjustment of constants, if necessary) you see that the numerator is the derivative of the denominator, then the integral is the logarithm of the modulus of the denominator.

You can use this rule to integrate all rational functions of x in which the denominator is a linear function, by a suitable adjustment of constants.

Example 11.8.1

$$\int \frac{dx}{ax} = \frac{1}{a} \int \frac{a}{ax} dx = \frac{1}{a} \ln|ax| + C$$

Example 11.8.2

$$\int \frac{dx}{ax+b} = \frac{1}{a} \ln|ax+b| + C$$

Example 11.8.3

$$\int \frac{x\,dx}{2x^2+3} = \frac{1}{4} \int \frac{4x\,dx}{2x^2+3} = \frac{1}{4} \ln|2x^2+3| + C$$

Example 11.8.4

$$\int \frac{2(x+1)dx}{x^2+2x+7} = \int \frac{(2x+2)dx}{x^2+2x+7} = \ln|x^2+2x+7| + C$$

Example 11.8.5

$$\int \tan x\,dx = \int \frac{\sin x}{\cos x} dx = -\int \frac{-\sin x}{\cos x} dx = -\ln|\cos x| + C$$

Example 11.8.6

$$\int \cot x\, dx = \int \frac{\cos x}{\sin x}\, dx = \ln|\sin x| + C$$

Example 11.8.7

$$\int \frac{6x+5}{3x^2+5x+2}\, dx = \ln\left|3x^2+5x+2\right| + C$$

Example 11.8.8

Integrate $\int (x+2)(2x-1)dx$

Although there is a rule for differentiating the product of two functions, there is no corresponding rule for integrating a product as in the above example. In such a case you have to multiply the factors.

$$\int (x+2)(2x-1)dx = \int (2x^2+3x-2)dx = \tfrac{2}{3}x^3 + \tfrac{3}{2}x^2 - 2x + C$$

Example 11.8.9

Integrate $\int \frac{x^4+3x^2+1}{x^3}\, dx$

In this example you can use a device which will be used later in more complicated cases. Split up the fraction into its component fractions.

$$\int \frac{x^4+3x^2+1}{x^3}\, dx = \int (x + 3x^{-1} + x^{-3})dx = \tfrac{1}{2}x^2 + 3\ln|x| - \frac{1}{2x^2} + C$$

Example 11.8.10

Given that $\frac{d^2y}{dx^2} = x^3$, find y

Since $\frac{d^2y}{dx^2}$ is the derivative of $\frac{dy}{dx}$, it follows that by integrating $\frac{d^2y}{dx^2}$ you obtain $\frac{dy}{dx}$. Having thus found $\frac{dy}{dx}$ integrating again will give the equation connecting y and x. From $\frac{d^2y}{dx^2} = x^3$ by integrating,

$$\frac{dy}{dx} = \int x^3\, dx = \tfrac{1}{4}x^4 + C_1$$

Integrating again, $y = \int\left(\frac{1}{4}x^4 + C_1\right)dx = \frac{1}{20}x^5 + C_1x + C_2$

Therefore $\qquad y = \frac{1}{20}x^5 + C_1x + C_2$

As a result of integrating twice, two constants, C_1 and C_2, are introduced. To find C_1 and C_2 it is necessary to have two pairs of corresponding values of x and y. On substituting these values of x and y you get two simultaneous equations involving C_1 and C_2. After solving them, substitute the values you find in the equation:

$$y = \frac{1}{20}x^5 + C_1x + C_2$$

and you have the equation relating x to y.

Nugget – Order, order

At the risk of stating the obvious, if you are differentiating or integrating a polynomial function, a check on the order of the answer (i.e. the value of its highest power) will help to ensure that you haven't done anything silly. Just as differentiation *reduces* the order of a polynomial function by 1, so integration does the reverse – it *increases* the order of a polynomial function by 1. Thus:

For differentiation:

▶ a quintic (of order 5) turns into a quartic (of order 4)

▶ a quartic (of order 4) turns into a cubic (of order 3)

▶ a cubic (of order 3) turns into a quadratic (of order 2)

▶ a quadratic (of order 2) turns into a linear (of order 1), and

▶ a linear (of order 1) turns into a constant number (of order 0)

For integration:

▶ a constant number (of order 0) turns into a linear (of order 1)

▶ a linear (of order 1) turns into a quadratic (of order 2)

▶ a quadratic (of order 2) turns into a cubic (of order 3)

▶ a cubic (of order 3) turns into a quartic (of order 4), and

▶ a quartic (of order 4) turns into a quintic (of order 5)

and so on.

Exercise 11.1

In questions 1 to 48 find the integrals.

1 $\int 3x\, dx$ **2** $\int 5x^2\, dx$ **3** $\int \frac{1}{2}x^3\, dx$

4 $\int 0.4x^4\, dx$ **5** $\int 12x^8\, dx$ **6** $\int 15t^2\, dt$

7 $\int \dfrac{dx}{2}$ **8** $\int d\theta$ **9** $\int (4x^2 - 5x + 1)dx$

10 $\int (3x^4 - 5x^3)dx$ **11** $\int x\{8x - \frac{1}{2}\}dx$

12 $\int 6x^2(x^2 + x)dx$ **13** $\int \{(x + 3)(x - 3)\}dx$

14 $\int \{(2x - 3)(x + 4)\}dx$ **15** $\int \dfrac{dx}{x^2}$ **16** $\int \dfrac{dx}{x^{1.4}}$

17 $\int 3x^{-4}dx$ **18** $\int \sqrt[3]{x}\,dx$ **19** $\int \frac{1}{2}x^{-\frac{1}{2}}dx$

20 $\int \dfrac{dx}{\sqrt[3]{x}}$ **21** $\int \left(x^{\frac{1}{2}} + x^{-\frac{1}{2}}\right)dx$

22 $\int \left(x^{\frac{2}{3}} + 1 + x^{-\frac{2}{3}}\right)dx$ **23** $\int \dfrac{1}{2\sqrt{2x^3}}\, dx$ **24** $\int \left(\frac{1}{2}\pi - 5x^{-0.5}\right)dx$

25 $\int g\, dt$ **26** $\int \left(\dfrac{1}{x^3} - \dfrac{1}{x^2} + \dfrac{1}{x} - 1\right)dx$

27 $\int \sqrt{t}\, dt$ **28** $\int \left(2 - \frac{1}{3}x^2 - \dfrac{1}{2\sqrt{x}}\right)dx$

29 $\int \dfrac{1.4}{u}\, dx$ **30** $\int \dfrac{dx}{u + 3}$ **31** $\int \dfrac{dx}{au + b}$

32 $\int \left(\dfrac{3}{x-1} - \dfrac{4}{x-2}\right)dx$ **33** $\int \dfrac{2x}{x^2 + 4}\, dx$ **34** $\int \dfrac{dx}{3 - 2x}$

35 $\int \dfrac{x + 3}{x}\, dx$ **36** $\int \dfrac{x^3 - 7}{x}\, dx$ **37** $\int \dfrac{x^2 - x + 1}{x^3}\, dx$

38 $\int \sqrt{ax + b}\, dx$ **39** $\int \sqrt{2x + 3}\, dx$ **40** $\int \sqrt{1 + \frac{x}{2}}\, dx$

41 $\int \dfrac{dx}{\sqrt{ax + b}}$ **42** $\int \dfrac{dx}{\sqrt{1 - x}}$ **43** $\int (ax + b)^2 dx$

44 $\int x(1 + x)(1 + x^2)dx$ **45** $\int \dfrac{x\, dx}{x^2 - 1}$ **46** $\int \dfrac{\sin ax\, dx}{1 + \cos ax}$

47 $\displaystyle\int \frac{e^{3x}dx}{e^{3x}+6}$ **48** $\displaystyle\int \frac{1+\cos 2x}{2x+\sin 2x}\,dx$

In questions 49 to 54, use the given information to solve the questions asked.

49 If $\dfrac{d^2y}{dx^2} = 3x^2$, find y in terms of x.

50 If $\dfrac{dy}{dx} = 6x^2$, find y in term of x, when $y = 5$ if $x = 1$.

51 If $\dfrac{d^2y}{dx^2} = 5x$, find y in terms of x when it is known that if $x = 2$, $\dfrac{dy}{dx} = 12$ and when $x = 1$, $y = 1$.

52 The gradient of a curve is given by $\dfrac{dy}{dx} = 4x - 5$. When $x = 1$ it is known that $y = 3$. Find the equation of the curve.

53 The gradient of a curve is given by $\dfrac{dy}{dx} = 9x^2 - 10x + 4$. If the curve passes through the point $(1, 6)$, find its equation.

54 If $\dfrac{d^2s}{dt^2} = 8t$, find s in terms of t, when it is known that if $t = 0$, $s = 10$ and $\dfrac{ds}{dt} = 8$.

11.9 Integrals of standard forms

There are a number of integrals known as **standard forms**. They are derivatives of functions that you already know.

ALGEBRAIC FUNCTIONS

1 $\displaystyle\int x^n\,dx = \frac{1}{n+1}x^{n+1} + C$

2 $\displaystyle\int \frac{dx}{x} = \ln|x| + C$

3 $\displaystyle\int a^x\,dx = \frac{1}{\ln a}a^x + C$

4 $\displaystyle\int e^x\,dx = e^x + C$

TRIGONOMETRIC FUNCTIONS

5 $\displaystyle\int \sin x \, dx = -\cos x + C$

6 $\displaystyle\int \cos x \, dx = \sin x + C$

7 $\displaystyle\int \tan x \, dx = -\ln|\cos x| + C = \ln|\sec x| + C$

8 $\displaystyle\int \cot x \, dx = \ln|\sin x| + C$

Note: The derivatives of $\sec x$ and $\csc x$ namely $\sec x \tan x$ and $-\csc x \cot x$, do not give rise to standard forms, but to products of them. They are not therefore included in the list above, but follow below. The integrals of $\sec x$ and $\csc x$ do not arise by direct differentiation. They will be given later in Section 12.5.

HYPERBOLIC FUNCTIONS

9 $\displaystyle\int \sinh x \, dx = \cosh x + C$

10 $\displaystyle\int \cosh x \, dx = \sinh x + C$

11 $\displaystyle\int \tanh x \, dx = \ln \cosh x + C$

12 $\displaystyle\int \coth x \, dx = \ln|\sinh x| + C$

Note the following variations (on the standard forms given above).

$$\int \sin ax \, dx = -\frac{1}{a} \cos ax + C$$

$$\int \sin(ax + b) \, dx = -\frac{1}{a} \cos(ax + b) + C$$

$$\int \cos ax \, dx = \frac{1}{a} \sin ax + C$$

$$\int \cos(ax + b) \, dx = \frac{1}{a} \sin(ax + b) + C$$

$$\int \tan ax \, dx = \frac{1}{a} \ln|\sec ax| + C$$

$$\int \sinh ax \, dx = \frac{1}{a} \cosh ax + C$$

Exercise 11.2

Find the following integrals.

1 $\int 3e^{2x}\, dx$

2 $\int e^{3x-1}\, dx^3$

3 $\int (e^x + e^{-x})^2\, dx$

4 $\int e^{x/a}\, dx$

5 $\int \left(e^{\frac{1}{2}x} + e^{-\frac{1}{2}x} \right) dx$

6 $\int (e^{ax} - e^{-ax})\, dx$

7 $\int (e^{3x} + a^{3x})\, dx$

8 $\int 2^x\, dx$

9 $\int 10^{3x}\, dx$

10 $\int (a^x + a^{-x})\, dx$

11 $\int xe^{x^2}\, dx$

12 $\int \sin x\, e^{\cos x}\, dx$

13 $\int \sin 3x\, dx$

14 $\int \cos 5x\, dx$

15 $\int \sin \frac{1}{2}\left(x + \frac{1}{3}\pi \right) dx$

16 $\int \cos(2x + a)\, dx$

17 $\int \sin \frac{1}{3}x\, dx$

18 $\int \sin(\alpha - 3x)\, dx$

19 $\int (\cos ax + \sin bx)\, dx$

20 $\int \sin 2ax\, dx$

21 $\int (\cos 3x - \sin \frac{1}{3}x)\, dx$

22 $\int \dfrac{1 + \cos x}{x + \sin x}\, dx$

23 $\int \sin^3 x \cos x\, dx$

24 $\int \sec^2 x\, e^{\tan x}\, dx$

25 $\int (\tan ax + \cot bx)\, dx$

26 $\int \dfrac{\sin 2x}{1 + \sin^2 x}\, dx$

27 $\int \cosh 2x\, dx$

28 $\int \sinh \frac{1}{3}ax\, dx$

29 $\int \tanh 3x\, dx$

30 $\int \sin(ax + b)\, dx$

31 $\int \dfrac{(e^x + 1)^2}{\sqrt{e^x}}\, dx$

32 $\int \tan \frac{3}{2}x\, dx$

33 $\int \sec^2 \frac{1}{3}x\, dx$

34 $\int \dfrac{e^x}{1 + e^x}\, dx$

35 $\int \dfrac{\sec^2 x}{1 + \tan x}\, dx$

36 $\int \cos x \sqrt{\sin x}\, dx$

11.10 Additional standard integrals

Here are some more standard forms which are useful for integration. Some of them are the results of differentiating standard forms.

TRIGONOMETRIC FUNCTIONS

13 $\int \sec x\, \tan x\, dx = \sec x + C$

14 $\int \operatorname{cosec} x\, \cot x\, dx = -\operatorname{cosec} x + C$

15 $\int \operatorname{cosec}^2 x\, dx = -\cot x + C$

16 $\int \sec^2 x\, dx = \tan x + C$

INVERSE TRIGONOMETRIC FUNCTIONS

17 $\int \dfrac{dx}{\sqrt{a^2 - x^2}} = \sin^{-1}\dfrac{x}{a} + C$

18 $\int \dfrac{dx}{a^2 + x^2} = \dfrac{1}{a}\tan^{-1}\dfrac{x}{a} + C \text{ or } -\dfrac{1}{a}\cot^{-1}\dfrac{x}{a} + C$

19 $\int \dfrac{dx}{x\sqrt{x^2 - a^2}} = \dfrac{1}{a}\sec^{-1}\dfrac{x}{a} + C \text{ or } -\dfrac{1}{a}\operatorname{cosec}^{-1}\dfrac{x}{a} + C$

20 $\int \dfrac{dx}{\sqrt{a^2 + x^2}} = \sinh^{-1}\left(\dfrac{x}{a}\right) + C$

21 $\int \dfrac{dx}{\sqrt{x^2 - a^2}} = \cosh^{-1}\left(\dfrac{x}{a}\right) + C$

22 $\int \dfrac{dx}{a^2 - x^2} = \dfrac{1}{a}\tanh^{-1}\left(\dfrac{x}{a}\right) + C, \quad x^2 < a^2$

23 $\int \dfrac{dx}{a^2 - x^2} = \dfrac{1}{a}\coth^{-1}\left(\dfrac{x}{a}\right) + C, \quad x^2 > a^2$

24 $\int \dfrac{dx}{x\sqrt{a^2 - x^2}} = -\dfrac{1}{a}\operatorname{sech}^{-1}\left(\dfrac{x}{a}\right) + C$

25 $\int \dfrac{dx}{x\sqrt{a^2 + x^2}} = -\dfrac{1}{a}\operatorname{cosech}^{-1}\left(\dfrac{x}{a}\right) + C$

Nugget – The 25 standard integrals

In this section you have been given a list of the 25 most useful standard integrals. These represent standard forms that you will need to look out for when attempting integrations in the future. Unfortunately, you cannot recognize a standard form unless it is familiar to you, so this is where your hard work at learning these templates will pay great dividends in the long run. Spend as long as it takes now to read and reread them, until you are familiar with all of them, and really know what you are looking for when tackling the practice examples that follow.

Example 11.10.1

Find, $I = \int \dfrac{dx}{\sqrt{16 - 9x^2}}$

You can transform this integral into standard form number 17. Try it before you look below.

$$I = \int \frac{dx}{\sqrt{16 - 9x^2}} = \int \frac{1}{3} \frac{dx}{\sqrt{\frac{16}{9} - x^2}} = \frac{1}{3} \int \frac{dx}{\sqrt{\frac{16}{9} - x^2}}$$

This is now in the form of number 17, with $a = \frac{4}{3}$

Therefore $\qquad I = \dfrac{1}{3} \sin^{-1}\left(\dfrac{x}{\frac{4}{3}}\right) + C = \dfrac{1}{3} \sin^{-1}\left(\dfrac{3x}{4}\right) + C$

Example 11.10.2

Find the integral $I = \int \dfrac{dx}{\sqrt{9x^2 - 1}}$

Compare this integral with standard form number 21.

$$I = \int \frac{dx}{\sqrt{9x^2 - 1}} = \int \frac{dx}{3\sqrt{x^2 - \frac{1}{9}}} = \frac{1}{3} \int \frac{dx}{\sqrt{x^2 - \frac{1}{9}}}$$

Therefore $\qquad I = \dfrac{1}{3} \cosh^{-1} 3x + C$ or $\dfrac{1}{3} \ln \left| 3x + \sqrt{9x^2 - 1} \right| + C$

Example 11.10.3

Find the integral $I = \int \dfrac{dx}{9x^2 + 4}$

This form is a variation of number 18.

$$I = \int \frac{dx}{9x^2 + 4} = \int \frac{dx}{9\left(x^2 + \frac{4}{9}\right)} = \frac{1}{9} \int \frac{dx}{\left(x^2 + \frac{4}{9}\right)}$$

Therefore $\qquad I = \dfrac{1}{9} \times \dfrac{3}{2} \tan^{-1}\left(\dfrac{x}{\frac{2}{3}}\right) + C = \dfrac{1}{6} \tan^{-1}\left(\dfrac{3x}{2}\right) + C$

Exercise 11.3

Find the following integrals.

1 $\int \dfrac{dx}{\sqrt{9-x^2}}$ **2** $\int \dfrac{dx}{\sqrt{x^2-9}}$ **3** $\int \dfrac{dx}{\sqrt{x^2+9}}$

4 $\int \dfrac{dx}{9+x^2}$ **5** $\int \dfrac{dx}{9-x^2}$ **6** $\int \dfrac{dx}{x^2-9}$

7 $\int \dfrac{dx}{\sqrt{16-x^2}}$ **8** $\int \dfrac{dx}{16-x^2}$ **9** $\int \dfrac{dx}{\sqrt{x^2-16}}$

10 $\int \dfrac{dx}{x^2-16}$ **11** $\int \dfrac{dx}{\sqrt{x^2+16}}$ **12** $\int \dfrac{dx}{x^2+16}$

13 $\int \dfrac{dx}{\sqrt{25-9x^2}}$ **14** $\int \dfrac{dx}{\sqrt{9x^2+25}}$ **15** $\int \dfrac{dx}{\sqrt{9x^2+25}}$

16 $\int \dfrac{dx}{4x^2+9}$ **17** $\int \dfrac{dx}{9-4x^2}$ **18** $\int \dfrac{dx}{4x^2-9}$

19 $\int \dfrac{dx}{9x^2+4}$ **20** $\int \dfrac{dx}{\sqrt{9x^2+4}}$ **21** $\int \dfrac{dx}{\sqrt{9x^2-4}}$

22 $\int \dfrac{dx}{\sqrt{49x^2+25}}$ **23** $\int \dfrac{dx}{\sqrt{2x^2+5}}$ **24** $\int \dfrac{dx}{\sqrt{5-x^2}}$

25 $\int \dfrac{dx}{\sqrt{5-4x^2}}$ **26** $\int \dfrac{dx}{\sqrt{4x^2+5}}$ **27** $\int \dfrac{dx}{25-4x^2}$

28 $\int \dfrac{dx}{\sqrt{7x^2+36}}$ **29** $\int \dfrac{dx}{x\sqrt{1+x^2}}$ **30** $\int \dfrac{dx}{x\sqrt{x^2-4}}$

31 $\int \dfrac{dx}{x\sqrt{x^2+4}}$ **32** $\int \dfrac{dx}{x\sqrt{4-x^2}}$

Key ideas

▶ Integration is the converse of differentiation. In general, if $f'(x)$ represents the differential coefficient of $f(x)$, then questions to do with integration are of the form, 'given $f'(x)$, find $f(x)$'.

▶ When a function is differentiated, any constant that it contains will disappear. Thus, when a function is integrated, the constant cannot be determined (without further information) and so an unknown constant, C, is added. This is called the *constant of integration*.

▶ An *indefinite integral* is one for which the constant term is unknown.

▶ You saw in Chapter 4 that the derivative of x^n with respect to x is $n \times x^{n-1}$. It follows from reversing this rule that the integral of x^n with respect to x is $\frac{1}{n+1} x^{n+1} + C$.

▶ Another useful rule for integration is that $\int f'(x)/f(x)\, dx = \log f(x) + C$.

▶ A summary of the integrals of standard forms is given in Section 11.9 and 11.10.

12

Methods of integration

In this chapter you will learn:

▶ *more systematic methods of integration*
▶ *the method of integration by substitution*
▶ *the method of integration by parts.*

12.1 Introduction

This chapter contains some of the rules and devices for integration which were referred to in Section 11.1. The general aim of these methods is not to integrate directly, but to transform the function to be integrated, so that it takes the form of one of the known standard integrals given at the end of the last chapter.

12.2 Trigonometric functions

You can often make use of trigonometric formulae to change products or powers of trigonometric functions into sums of other functions. You can then use the rules of Section 11.6 or Section 11.8 to carry out the integration. This process was used in Examples 11.8.5 and 11.8.6, where, by changing $\tan x$ to $\frac{\sin x}{\cos x}$ and $\cot x$ to $\frac{\cos x}{\sin x}$, the integrals of $\tan x$ and $\cot x$ were found.

Example 12.2.1

Among the formulae which are commonly used are the following two

$$\sin^2 x = \tfrac{1}{2}\left(1 - \cos 2x\right)$$

and
$$\cos^2 x = \tfrac{1}{2}\left(1 + \cos 2x\right)$$

Hence
$$\int \sin^2 x\, dx = \int \tfrac{1}{2}\left(1 - \cos 2x\right) dx$$
$$= \tfrac{1}{2}\int \left(1 - \cos 2x\right) dx$$
$$= \tfrac{1}{2}\left(x - \tfrac{1}{2}\sin 2x\right) + C$$

Similarly,
$$\int \cos^2 x\, dx = \tfrac{1}{2}\left(x + \tfrac{1}{2}\sin 2x\right) + C$$

Notice that in each case the formula used enabled you to change a power of the function into a sum, after which you could integrate it. Here are some more examples.

Example 12.2.2

Using
$$\tan^2 x = \sec^2 x - 1$$

$$\int \tan^2 x \, dx = \int (\sec^2 x - 1) \, dx = \tan x - x + C$$

Similarly
$$\int \cot^2 x \, dx = \int (\operatorname{cosec}^2 x - 1) \, dx = -\cot x - x + C$$

Example 12.2.3

Find $\int \sin^3 x \, dx$

You can find this integral by using the formula

$$\sin 3A = 3 \sin A - 4 \sin^3 A$$

in the form
$$\sin^3 A = \tfrac{1}{4}(3 \sin A - \sin 3A)$$

$$\int \sin^3 x \, dx = \int \left(\tfrac{3}{4} \sin x - \tfrac{1}{4} \sin 3x \right) dx = -\tfrac{3}{4} \cos x + \tfrac{1}{12} \cos 3x + C$$

You can use the formula $\cos 3A = 4 \cos^3 A - 3 \cos A$ in a similar way to find $\int \cos^3 x \, dx$.

The following formulae are useful for changing products of sines and cosines into sums of these functions:

$$\sin A \cos B = \tfrac{1}{2} \{ \sin(A + B) + \sin(A - B) \}$$
$$\cos A \sin B = \tfrac{1}{2} \{ \sin(A + B) - \sin(A - B) \}$$
$$\cos A \cos B = \tfrac{1}{2} \{ \cos(A + B) + \cos(A - B) \}$$
$$\sin A \sin B = \tfrac{1}{2} \{ \cos(A - B) - \cos(A + B) \}$$

Example 12.2.4

Find the integral $\int \dfrac{\sin^3 x}{\cos^2 x} \, dx$

Rearranging $\int \dfrac{\sin^3 x}{\cos^2 x} \, dx = \int \dfrac{\sin^2 x \sin x}{\cos^2 x} \, dx$

$$= \int \frac{\left(1 - \cos^2 x\right)\sin x}{\cos^2 x} dx$$

$$= \int \frac{\sin x}{\cos^2 x} dx - \int \frac{\cos^2 x \sin x}{\cos^2 x} dx$$

$$= \int \sec x \tan x \, dx - \int \sin x \, dx$$

$$= \sec x + \cos x + C$$

Example 12.2.5

Integrate $\int \sin 3x \cos 4x \, dx$

Using the second formula at the end of Example 12.2.3 (see above)

$$\int \cos 4x \sin 3x \, dx = \int \tfrac{1}{2}\left(\sin(4x + 3x) - \sin(4x - 3x)\right)dx$$

$$= \tfrac{1}{2}\int \sin 7x \, dx - \tfrac{1}{2}\int \sin x \, dx$$

$$= -\tfrac{1}{2} \times \tfrac{1}{7} \cos 7x + \tfrac{1}{2} \cos x + C$$

Exercise 12.1

Find the following integrals.

1 $\int \sin^2 \tfrac{1}{2} x \, dx$ **2** $\int \cos^2 \tfrac{1}{2} x \, dx$ **3** $\int \tan^2 \tfrac{1}{2} x \, dx$

4 $\int \cos^4 x \, dx$ **5** $\int \sin^4 x \, dx$ **6** $\int \cot^2 2x \, dx$

7 $\int \sin^2 2x \, dx$ **8** $\int \cos^2 3x \, dx$ **9** $\int \cos^2 (ax + b) dx$

10 $\int \sin^3 x \, dx$ **11** $\int \cos^3 x \, dx$ **12** $\int \sin 2x \sin 3x \, dx$

13 $\int \cos 3x \cos x \, dx$ **14** $\int \sin 4x \cos 2x \, dx$ **15** $\int \sin 4x \cos \tfrac{3}{2} x \, dx$

16 $\int \sin^2 x \cos^2 x \, dx$ **17** $\int \dfrac{dx}{\sin^2 x \cos^2 x}$ **18** $\int \dfrac{1 + \sin^2 x}{\cos^2 x} dx$

19 $\int \tan^3 x \, dx$ **20** $\int \sqrt{1 + \cos x} \, dx$

12.3 Integration by substitution

You can sometimes, by changing the independent variable, transform a function into a form which you can integrate. Experience will suggest the particular form of substitution which is likely to be effective, but there are some recognizable forms in which you can use standard substitutions. You can often treat irrational functions in this way, as you will see in the following examples.

Nugget – The four stages of Integration by substitution

All examples of integration by substitution can be reduced to the following four stages:

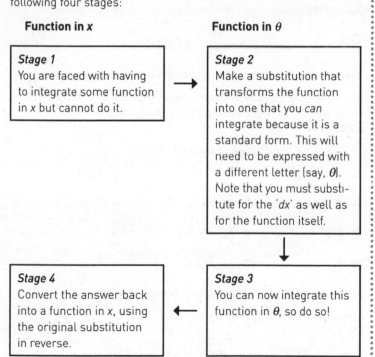

Function in x	Function in θ

Stage 1
You are faced with having to integrate some function in x but cannot do it.

Stage 2
Make a substitution that transforms the function into one that you *can* integrate because it is a standard form. This will need to be expressed with a different letter (say, θ). Note that you must substitute for the 'dx' as well as for the function itself.

Stage 4
Convert the answer back into a function in x, using the original substitution in reverse.

Stage 3
You can now integrate this function in θ, so do so!

12.4 Some trigonometrical substitutions

Consider the integral $\int \sqrt{a^2 - x^2}\, dx$

The treatment that follows gives the theory for finding this integral, but writing the solution is usually done differently. So immediately after the theory there is another solution which, in practice, is the one usually used for evaluating the integral.

If you think about the meaning of integration, you will see that if $y = \int \sqrt{a^2 - x^2}\, dx$, then you are looking for a function such that $\frac{dy}{dx} = \sqrt{a^2 - x^2}$. The form of this expression suggests that if you replace x by $a\sin\theta$ you get $a^2 - a^2\sin^2\theta$, that is, $a^2(1 - \sin^2\theta)$. This is equal to $a^2\cos^2\theta$.

Taking the square root, the irrational quantity disappears.

Thus
$$\frac{dy}{dx} = a\cos\theta$$

You are then left with two independent variables, namely, x and θ.

Remembering the function of a function rule $\frac{dy}{d\theta} = \frac{dy}{dx} \times \frac{dx}{d\theta}$, and using the fact that if $x = a\sin\theta$ then $\frac{dx}{d\theta} = a\cos\theta$

$$\frac{dy}{d\theta} = a\cos\theta \times a\cos\theta$$

so
$$y = \int a^2\cos^2\theta\, d\theta$$

Therefore
$$y = a^2\left(\tfrac{1}{2}\left(\theta + \tfrac{1}{2}\sin 2\theta\right)\right) + C$$
$$= \tfrac{1}{2}a^2\theta + \tfrac{1}{2}a\sin\theta \times a\cos\theta + C$$

You must now change the variable from θ back to x.

Since $x = a\sin\theta$ and $\sin\theta = \frac{x}{a}$, $\theta = \sin^{-1}\frac{x}{a}$

Also
$$a\cos\theta = a\sqrt{1 - \sin^2\theta} = \sqrt{a^2 - a^2\sin^2\theta}$$
$$= \sqrt{a^2 - x^2}$$

Substituting in the expression for y in terms of θ

$$y = \tfrac{1}{2} a^2 \sin^{-1}\tfrac{x}{a} + \tfrac{1}{2} x \sqrt{a^2 - x^2} + C$$

so $$\int \sqrt{a^2 - x^2} \, dx = \frac{1}{2} a^2 \sin^{-1} \frac{x}{a} + \frac{1}{2} x \sqrt{a^2 - x^2} + C$$

Here, in practice, is how the solution would be written.

To find $\int \sqrt{a^2 - x^2} \, dx$, make the substitution $x = a\sin\theta$.

Then, instead of writing $\frac{dx}{d\theta} = a\cos\theta$, you write $dx = a\cos\theta \, d\theta$ and think in terms of substituting for dx in the integral.

Thus $$\int \sqrt{a^2 - x^2} \, dx = \int \sqrt{a^2 - a^2 \sin^2\theta} \times a\cos\theta \, d\theta$$

$$= \int a\cos\theta \times a\cos\theta \, d\theta = \int a^2 \cos^2\theta \, d\theta$$

The solution then proceeds as before.

Example 12.4.1

Find $\int \dfrac{dx}{\sqrt{a^2 - x^2}}$

Using the same substitution, $x = a\sin\theta$, you find:

$$\sqrt{a^2 - x^2} = a\cos\theta$$

and $$dx = a\cos\theta \, d\theta$$

so $$\int \frac{dx}{\sqrt{a^2 - x^2}} = \int \frac{a\cos\theta \, d\theta}{a\cos\theta}$$

$$= \int d\theta = \theta + C = \sin^{-1}\frac{x}{a} + C$$

Example 12.4.2

Find $\int \dfrac{dx}{a^2 + x^2}$

The form of the integral suggests the substitution $x = a\tan\theta$, as

$$a^2 + x^2 = a^2 + a^2 \tan^2\theta$$

$$= a^2 \left(1 + \tan^2\theta\right)$$

$$= a^2 \sec^2\theta$$

So using $x = a \tan\theta$ gives $dx = a \sec^2 \theta \, d\theta$

Therefore
$$\int \frac{dx}{a^2 + x^2} = \int \frac{a \sec^2 \theta \, d\theta}{a^2 \sec^2 \theta}$$

$$= \frac{1}{a} \int d\theta$$

$$= \frac{1}{a} \theta + C$$

$$= \frac{1}{a} \tan^{-1} \frac{x}{a} + C$$

12.5 The substitution $t = \tan \frac{1}{2} x$

An occasionally useful trigonometrical substitution is given by the following formulae, in which $\sin x$ and $\cos x$ are expressed in terms of $\tan \frac{1}{2} x$.

$$\sin x = \frac{2 \tan \frac{1}{2} x}{1 + \tan^2 \frac{1}{2} x} \qquad \cos x = \frac{1 - \tan^2 \frac{1}{2} x}{1 + \tan \frac{1}{2} x}$$

When using these formulae it is convenient to proceed as follows.

Let $t = \tan \frac{1}{2} x$ Then $\sin x = \frac{2t}{1 + t^2}$ and $\cos x = \frac{1 - t^2}{1 + t^2}$

Since
$$t = \tan \frac{1}{2} x$$

$$dt = \frac{1}{2} \sec^2 \frac{1}{2} x \, dx$$

so
$$dx = \frac{2 \, dt}{\sec^2 \frac{1}{2} x} = \frac{2 \, dt}{1 + t^2}$$

You can use this substitution to find the following integral.

Example 12.5.2

Find $\int \sec x \, dx$

Using the fact that $\cos x = \sin\left(x + \frac{1}{2}\pi\right)$, you can write the integral of $\sec x$ in the form

$$\int \sec x \, dx = \int \operatorname{cosec}\left(x + \frac{1}{2}\pi\right) dx$$

Therefore

$$\int \sec x \, dx = \ln\left|\tan\left(\tfrac{1}{2}x + \tfrac{1}{4}\pi\right)\right| + C$$

or

$$\int \sec x \, dx = \ln\left|A \tan\left(\tfrac{1}{2}x + \tfrac{1}{4}\pi\right)\right|$$

You can also show that $\int \sec x \, dx = \ln|A(\sec x + \tan x)|$

Example 12.5.3

Find $\int \dfrac{dx}{5 + 4\cos x}$

Let $t = \tan \frac{1}{2}x$ so that $dx = \dfrac{2\,dt}{1 + t^2}$ and $\cos x = \dfrac{1 - t^2}{1 + t^2}$

Then

$$\int \frac{dx}{5 + 4\cos x} = \int \frac{2\,dt}{1 + t^2} \div \frac{1}{5 + 4\dfrac{1 - t^2}{1 + t^2}}$$

$$= \int \frac{2\,dt}{5\left(1 + t^2\right) + 4\left(1 - t^2\right)} = \int \frac{2\,dt}{9 + t^2}$$

Therefore
$$\int \frac{dx}{5+4\cos x} = \frac{2}{3}\tan^{-1}\frac{t}{3} + C$$
$$= \frac{2}{3}\tan^{-1}\left(\frac{1}{3}\tan\frac{x}{2}\right) + C$$

In general the substitution $t = \tan\frac{1}{2}x$ is useful for calculating
$\int \dfrac{dx}{a+b\cos x}$ or $\int \dfrac{dx}{a+b\sin x}$ where a and b are constants.
The form of the result will depend on the relative values of a and b, and it may need methods which you will find in the next chapter.

12.6 Worked examples

Example 12.6.1
Integrate $\int \sqrt{16-9x^2}\ dx$

Let $3x = 4\sin\theta$

Then $\qquad x = \frac{4}{3}\sin\theta, \theta = \sin^{-1}\frac{3}{4}x, dx = \frac{4}{3}\cos\theta\ d\theta$

and $\qquad \cos\theta = \sqrt{1-\sin^2\theta} = \sqrt{1-\left(\frac{3}{4}x\right)^2} = \frac{1}{4}\sqrt{16-9x^2}$

Substituting these expressions in $\int \sqrt{16-9x^2}dx$

$$\int \sqrt{16-9x^2}\ dx = \int \sqrt{16-16\sin^2\theta} \times \frac{4}{3}\cos\theta\ d\theta$$
$$= 4 \times \frac{4}{3}\int \cos\theta \times \cos\theta\ d\theta$$
$$= \frac{16}{3}\int \cos^2\theta\ d\theta$$
$$= \frac{16}{3}\int \frac{1+\cos 2\theta}{2}d\theta$$
$$= \frac{8}{3}\left(\theta + \frac{1}{2}\sin 2\theta\right) + C$$
$$= \frac{8}{3}\left(\sin^{-1}\frac{3}{4}x + \sin\theta\cos\theta\right) + C$$
$$= \frac{8}{3}\left(\sin^{-1}\frac{3}{4}x + \frac{3}{4}x \times \frac{1}{4}\sqrt{16-9x^2}\right) + C$$
$$= \frac{8}{3}\sin^{-1}\frac{3}{4}x + \frac{1}{2}x\sqrt{16-9x^2} + C$$

Example 12.6.2

Integrate $\int \dfrac{dx}{\sqrt{2-3x^2}}$

Let $x = \sqrt{\tfrac{2}{3}}\sin\theta$

Then $dx = \sqrt{\tfrac{2}{3}}\cos\theta\, d\theta$ and $\cos\theta = \sqrt{1-\tfrac{3}{2}x^2} = \tfrac{1}{\sqrt{2}}\sqrt{2-3x^2}$

$$\int \frac{dx}{\sqrt{2-3x^2}} = \int \frac{\sqrt{\tfrac{2}{3}}\cos\theta\, d\theta}{\sqrt{2-2\sin^2\theta}}$$

$$= \int \frac{\sqrt{\tfrac{2}{3}}\cos\theta\, d\theta}{\sqrt{2}\cos\theta}$$

$$= \frac{1}{\sqrt{3}}\int d\theta = \frac{1}{\sqrt{3}}\theta + C$$

$$= \frac{1}{\sqrt{3}}\sin^{-1}\sqrt{\tfrac{3}{2}}x + C$$

Example 12.6.3

Integrate $\int \dfrac{dx}{x^2\sqrt{1+x^2}}$

Let $x = \tan\theta$

Then $\qquad dx = \sec^2\theta\, d\theta$ and $\sqrt{1+x^2} = \sec\theta$

$$\int \frac{dx}{x^2\sqrt{1+x^2}} = \int \frac{\sec^2\theta\, d\theta}{\tan^2\theta\sec\theta}$$

$$= \int \frac{1}{\cos\theta} \times \frac{\cos^2\theta}{\sin^2\theta}\, d\theta$$

$$= \int \frac{\cos\theta}{\sin^2\theta}\, d\theta$$

$$= -\frac{1}{\sin\theta} + C \quad \begin{pmatrix} \text{By inspection or} \\ \text{putting } \sin\theta = z \end{pmatrix}$$

$$= -\frac{\sec\theta}{\tan\theta} + C$$

$$= -\frac{\sqrt{1+x^2}}{x} + C$$

Exercise 12.2

Find the following integrals.

1 $\int \sqrt{25-x^2}\, dx$ **2** $\int \sqrt{9-x^2}dx$ **3** $\int \sqrt{1-4x^2}dx$

4 $\int \sqrt{9-4x^2}dx$ **5** $\int \dfrac{dx}{x^2\sqrt{1-x^2}}$ **6** $\int \dfrac{dx}{x^2\sqrt{a^2+x^2}}$

7 $\int \operatorname{cosec}\tfrac{1}{2}x\, dx$ **8** $\int \sec\tfrac{1}{2}x\, dx$ **9** $\int \operatorname{cosec} 3x\, dx$

10 $\int \sec x \operatorname{cosec} x\, dx$ **11** $\int \dfrac{dx}{1+\cos x}$ **12** $\int \dfrac{dx}{1+\sin x}$

13 $\int \dfrac{dx}{1-\sin x}$ **14** $\int (\sec x + \tan x)dx$ **15** $\int \dfrac{dx}{5+3\cos x}$

16 $\int \dfrac{dx}{5-3\cos x}$

12.7 Algebraic substitutions

Sometimes you can transform a function into a form you can
integrate by using an algebraic substitution which changes the
independent variable. The forms taken will depend upon the
kind of function to be integrated and, in general, experience
and experiment will guide you. Your general aim should be to
simplify the function so that it becomes easier to integrate.

A common example of this method involves square roots in which
the expression under the root sign is of the form $ax + b$. You can
integrate these forms by using the substitution $ax + b = u^2$.

Example 12.7.1

Integrate $\int x\sqrt{2x+1}dx$

Let $\quad u^2 = 2x + 1$ or $u = \sqrt{2x+1}$, so $x = \tfrac{1}{2}(u^2 - 1)$ and $dx = u\, du$

Then $\quad \displaystyle\int x\sqrt{2x+1}\, dx = \int \tfrac{1}{2}(u^2-1)\times u \times u\, du$

$$= \tfrac{1}{2}\int u^2(u^2-1)du$$

$$= \tfrac{1}{2}\int (u^4 - u^2)du$$

$$= \tfrac{1}{2}\left(\frac{u^5}{5} - \frac{u^3}{3}\right) + C$$

$$= \frac{1}{30}\left(3u^5 - 5u^3\right) + C$$

$$= \frac{1}{30}\left(3(2x+1)^{\frac{5}{2}} - 5(2x+1)^{\frac{3}{2}}\right) + C$$

Example 12.7.2

Integrate $\int \dfrac{x\,dx}{\sqrt{x+3}}$

Let $u^2 = x + 3$ or $u = \sqrt{x+3}$, so $x = u^2 - 3$ and $dx = 2u\,du$

Then
$$\int \frac{x\,dx}{\sqrt{x+3}} = \int \frac{(u^2 - 3) \times 2u\,du}{u}$$

$$= 2\int (u^2 - 3)\,du$$

$$= 2\left(\frac{u^3}{3} - 3u\right) + C$$

$$= \tfrac{2}{3}u(u^2 - 9) + C$$

$$= \tfrac{2}{3}(x - 6)\sqrt{x+3} + C$$

Example 12.7.3

Integrate $\int x^3\sqrt{1 - x^2}\,dx$

You might think from the form of this function that it would be useful to substitute $x = \sin\theta$, but if you try it you will find that you obtain an awkward product to integrate.

Try $z = \sqrt{1 - x^2}$, so $z^2 = 1 - x^2$ and $2z\,dz = -2x\,dx$

Then
$$\int x^3\sqrt{1 - x^2}\,dx = \int \left(1 - z^2\right)^{\frac{3}{2}} \times z \times \frac{-z\,dz}{\sqrt{1 - z^2}}$$

$$= -\int z^2\left(1 - z^2\right)dz$$

$$= -\left(\frac{z^3}{3} - \frac{z^5}{5}\right) + C$$

$$= -\frac{1}{15}z^3\left(5 - 3z^2\right) + C$$

$$= -\frac{1}{15}\left(1 - x^2\right)^{\frac{3}{2}}\left(2 + 3x^2\right) + C$$

Example 12.7.4

Integrate $\int \sin^3\theta \cos^2\theta\, d\theta$

If you tried the substitution $x = \sin\theta$ in the previous example, this is the form you would have obtained. In products such as this, note which trigonometric function appears raised to an odd power and then substitute for the other function.

Let $x = \cos\theta$ so $dx = -\sin\theta\, d\theta$ and $\sin\theta = \sqrt{1-x^2}$

Then
$$\sin^3\theta \cos^2\theta\, d\theta = \int \sin^2\theta \times \cos^2\theta \times \sin\theta\, d\theta$$
$$= \int (1-x^2) \times x^2 (-dx)$$
$$= -\int (x^2 - x^4)\, dx$$
$$= -\frac{x^3}{3} + \frac{x^5}{5} + C$$
$$= \tfrac{1}{15} x^3 (3x^2 - 5) + C$$
$$= \tfrac{1}{15} \cos^3\theta (3\cos^2\theta - 5) + C$$

Exercise 12.3

Find the following integrals.

1 $\int x^2 \cos x^3\, dx$ Put $x^3 = u$

2 $\int \dfrac{x^2 dx}{1-2x^3}$ Put $1 - 2x^3 = u$

3 $\int \dfrac{x\, dx}{\sqrt{1+x^2}}$

4 $\int \dfrac{dx}{\sqrt{2-5x}}$

5 $\int \dfrac{1}{\sqrt{x}} \sin\sqrt{x}\, dx$

6 $\int \dfrac{x^2 dx}{\sqrt{1+x^3}}$

7 $\int \dfrac{\sin x\, dx}{1+2\cos x}$

8 $\int \dfrac{\ln x\, dx}{x}$

9 $\int x\sqrt{5+x^2}\, dx$

10 $\int \dfrac{2x\, dx}{1+x^4}$

11 $\int x(x-2)^4\, dx$

12 $\int \dfrac{x^2\, dx}{(x+1)^3}$

13 $\int \dfrac{x\, dx}{\sqrt{x-1}}$

14 $\int x\sqrt{x-1}\, dx$

15 $\int \dfrac{x\, dx}{\sqrt{5-x^2}}$

16 $\int \dfrac{x^3 dx}{\sqrt{x^2-1}}$

17 $\int x^3 \sqrt{x^2-2}\, dx$

18 $\int \dfrac{dx}{\sqrt{x-3}}$

19 $\int \sin^3 x \cos^2 x\, dx$

20 $\int \sin^2 x \cos^5 x\, dx$

12.8 Integration by parts

This method for integration comes from the rule for differentiating a product in Section 6.2, namely:

$$\frac{d}{dx}(uv) = u\frac{dv}{dx} + v\frac{du}{dx}$$

in which u and v are functions of x.

Integrating throughout with respect to x, you get:

$$uv = \int u\frac{dv}{dx}dx + \int v\frac{du}{dx}dx$$

Since u and v are functions of x, you can write this more conveniently in the form:

$$uv = \int u\,dv + \int v\,du$$

If you know either of the integrals on the right-hand side, you can find the other. You thus have the choice of calculating either of two integrals, whichever is possible or easier. If, for example, you decide that you can find $\int v\,du$ you can then find the other integral, namely $\int u\,dv$ from the equation:

$$\int u\,dv = uv - \int v\,du$$

Here is an example to illustrate the method.

Suppose that you have to find $\int x\cos x\,dx$

Let $u = x$ and $dv = \cos x\,dx$. Then $du = dx$, and since $dv = \cos x\,dx$, $v = \int \cos x\,dx = \sin x$. (Ignore the constant of integration for the moment.)

Substituting in the formula:

$$\int u\,dv = uv - \int v\,du$$

you get

$$\int x\cos x\,dx = x\sin x - \int \sin x\,dx$$

Thus instead of finding the original integral, you now have to find the simpler one $\int \sin x\,dx$, which you know to be $\cos x$.

Therefore $\qquad \int x\cos x\,dx = x\sin x + \cos x$

It is at this stage you should put in the constant of integration. You could try yourself to see what happens if you put it in any earlier. It makes no difference.

Therefore $\qquad \int x \cos x\, dx = x \sin x + \cos x + C$

If you had chosen u and v the other way round, so that $u = \cos x$ and $dv = x\, dx$, then $du = -\sin x\, dx$ and $v = \frac{1}{2}x^2$.

Then, substituting in the formula:

$$\int u\, dv = uv - \int v\, du$$

you find that

$$\int x \cos x\, dx = \tfrac{1}{2}x^2 \cos x - \int \tfrac{1}{2}x^2 \times (-\sin x)\, dx$$

This integral is more difficult than the original, so this choice of u and v is not effective.

Example 12.8.1

Integrate $\int \ln x\, dx$

Since $\ln x$ produces a simple expression when you differentiate it, put $u = \ln x$ and $dv = dx$.

Therefore $u = \ln x$, so $du = \frac{1}{x}\, dx$, and $dv = dx$ so $v = x$.

Therefore, substituting in $\int u\, dv = uv - \int v\, du$, you find that

$$\int \ln x\, dx = \ln|x| \times x - \int x \times \frac{1}{x}\, dx$$
$$= x \ln|x| - \int dx$$
$$= x \ln|x| - x + C$$

Thus $\qquad \int \ln x\, dx = x \ln|x| - x + C$

This is an important result and well worth remembering.

Example 12.8.2

Integrate $\int x e^{ax}\, dx$

You know that e^{ax} produces the same result, except for constants, when you differentiate or integrate it, but the derivative of x is simple.

So let $u = x$, so $du = dx$, and $dv = e^{ax}\,dx$ so $v = \frac{1}{a}e^{ax}$

Substituting in $\int u\,dv = uv - \int v\,du$

$$\int xe^{ax}\,dx = x \times \frac{1}{a}e^{ax} - \int \frac{1}{a}e^{ax}\,dx$$

$$= \frac{1}{a}xe^{ax} - \frac{1}{a} \times \frac{1}{a}e^{ax} + C$$

$$= \frac{1}{a}xe^{ax} - \frac{1}{a^2}e^{ax} + C$$

Example 12.8.3

Integrate $\int x^2 \sin x\,dx$

Let $u = x^2$, so $du = 2x\,dx$ and $dv = \sin x\,dx$ so $v = -\cos x$

Substituting in $\int u\,dv = uv - \int v\,du$, you find

$$\int x^2 \sin x\,dx = x^2 \times (-\cos x) - \int (-\cos x) \times 2x\,dx$$

$$= -x^2 \cos x + 2\int x \cos x\,dx$$

You arrive at an integral which you cannot integrate by inspection, but you can recognize it as one which you can also integrate by parts.

This example was integrated at the beginning of Section 12.8 as

$$\int x \cos x\,dx = x \sin x + \cos x$$

Therefore $\int x^2 \sin x\,dx = -x^2 \cos x + 2x \sin x + 2\cos x + C$

This kind of repetition of integration by parts occurs frequently.

Example 12.8.4

Integrate $\int \sin^{-1} x\,dx$

Let $u = \sin^{-1} x$, so $du = \frac{x}{\sqrt{1-x^2}}\,dx$, and $dv = dx$ so $v = x$

Substituting in $\int u\,dv = uv - \int v\,du$, you find

$$\int \sin^{-1} x\,dx = x \sin^{-1} x - \int \frac{x}{\sqrt{1-x^2}}\,dx$$

In the new integral, the derivative of $(1 - x^2)$, or a numerical multiple of it, appears in the numerator, so

$$\int \frac{x}{\sqrt{1-x^2}}\, dx = -\sqrt{1-x^2}$$

Therefore $\int \sin^{-1} x\, dx = x \sin^{-1} x + \sqrt{1-x^2} + C$

Example 12.8.5

Integrate $\int e^x \cos x\, dx$

Let $u = e^x$, so $du = e^x\, dx$

and $dv = \cos x\, dx$ so $v = \sin x$

Substituting in $\int u\, dv = uv - \int v\, du$, you find

$$\int e^x \cos x\, dx = e^x \sin x - \int e^x \sin x\, dx$$

The new integral is of the same type as the original one; if you integrate by parts again you find that:

$$\int e^x \sin x\, dx = -e^x \cos x + \int e^x \cos x\, dx$$

Substituting this into the first integral:

$$\int e^x \cos x\, dx = e^x \sin x + e^x \cos x - \int e^x \cos x\, dx$$

Therefore $\qquad 2\int e^x \cos x\, dx = e^x \sin x + e^x \cos x$

so $\qquad \int e^x \cos x\, dx = \tfrac{1}{2} e^x \sin x + \tfrac{1}{2} e^x \cos x + C$

You can find the integrals $\int e^{ax} \cos(bx + x)dx$ and $\int e^{ax} \sin(bx + x)dx$ by a similar method.

Nugget – Integration by parts: two tips

Here are a couple of general tips about using integration by parts.

1 It is one thing to develop your skills at being able to integrate by parts to integrate the product of functions, but is it always the best method to choose in this situation? Before rushing into integration by parts, first see if you can find a suitable algebraic substitution – if this works, you will probably find it to be the easier method.

2 When using integration by parts (to integrate the product of functions), you always need to integrate one function and differentiate the other. Since integration is often harder than differentiation, make sure that you choose the easier one to integrate!

Exercise 12.4

Find the following integrals.

1 $\int x \sin x \, dx$ **2** $\int x \sin 3x \, dx$ **3** $\int x^2 \cos x \, dx$

4 $\int x \ln x \, dx$ **5** $\int x e^x \, dx$ **6** $\int x^2 e^x \, dx$

7 $\int x e^{-ax} \, dx$ **8** $\int e^x \cos 2x \, dx$ **9** $\int \cos^{-1} x \, dx$

10 $\int \tan^{-1} x \, dx$ **11** $\int x \tan^{-1} x \, dx$ **12** $\int e^x \sin x \, dx$

13 $\int x \sin x \cos x \, dx$ **14** $\int x \sec^2 x \, dx$

Key ideas

▶ This chapter shows how certain functions can be transformed so as to make them easier to integrate.

▶ Standard substitutions for integrals that relate to trigonometric functions are given in Sections 12.4 and 12.5.

▶ Useful algebraic substitutions are given in Section 12.7.

▶ A standard method for integrating certain awkward functions is integration by parts (see Section 12.8). The rule derives from the rule for differentiating the product of two functions.

▶ Remember that it is usually easier to differentiate a function than integrate it so when you have performed an integral, it is always a good idea to check it by differentiation.

13

Integration of algebraic fractions

In this chapter you will learn:

▶ *how to integrate functions which are algebraic fractions with quadratic denominators*

▶ *how to integrate algebraic fractions when the denominator is the square root of a quadratic function.*

13.1 Rational fractions

Certain types of fractions have occurred frequently among the functions which have been integrated in previous work. One of the most common is that in which the numerator can be expressed as the derivative of the denominator. You saw in Section 11.7 that:

$$\int \frac{f'(x)}{f(x)}\, dx = \ln |f(x)| + C$$

A special form of this will often appear in the work that follows where the denominator is of the form $ax + b$.

$$\int \frac{dx}{ax+b} = \frac{1}{a} \int \frac{a\, dx}{ax+b} = \frac{1}{a} \ln |ax+b| + C$$

Nugget – Check by differentiating

Don't just accept this last result as another piece of mathematical magic. You can quickly check this important result by differentiating the answer to confirm that you get back to the starting function $\frac{1}{ax+b}$

Variants of this form include fractions in which the numerator is of the same or higher degree than the denominator, simple examples of which have already occurred. You can often rewrite this type of fraction so that you can use the general rule quoted above. The worked examples that follow illustrate this point.

Example 13.1.1

Find $\int \frac{x^2}{x+1}\, dx$

The process is to divide the denominator into the numerator:

$$\frac{x^2}{x+1} = x - 1 + \frac{1}{x+1}$$

(You may need to look up this division process in an algebra book.)

$$\int \frac{x^2}{x+1}\,dx = \int \left(x-1+\frac{1}{x+1}\right)dx$$

$$= \frac{1}{2}x^2 - x + \ln|x+1| + C$$

Example 13.1.2

Evaluate $\int \dfrac{3x+1}{2x-3}\,dx$

$$\int \frac{3x+1}{2x-3}\,dx = \int \left(\frac{3}{2} + \frac{\frac{11}{2}}{2x-3}\right)dx$$

$$= \frac{3}{2}x + \frac{11}{4}\ln|2x-3| + C$$

Exercise 13.1

Find the following integrals.

1 $\int \dfrac{x\,dx}{x+2}$ **2** $\int \dfrac{x\,dx}{1-x}$

3 $\int \dfrac{x\,dx}{a+bx}$ **4** $\int \dfrac{x+1}{x-1}\,dx$

5 $\int \dfrac{1-x}{1+x}\,dx$ **6** $\int \dfrac{2x-1}{2x+3}\,dx$

7 $\int \dfrac{x^2}{x+2}\,dx$ **8** $\int \dfrac{x^2}{1-x}\,dx$

13.2 Denominators of the form $ax^2 + bx + c$

In the next sections, integrals of this form are considered:

$$\int \frac{(Ax+B)\,dx}{px^2+qx+r}$$

The crucial part is the quadratic form in the denominator. You can write any quadratic function in one of three ways. It can be a perfect square, the difference of two squares or the sum of two squares. Which it is determines the form of the integral. Here are some examples.

Example 13.2.1

1 $x^2 + 4x + 4 = (x + 2)^2$ is a **perfect square**.

2 $x^2 + 4x + 2 = x^2 + 4x + (2)^2 - 2$

$$= (x + 2)^2 - 2 = (x + 2)^2 - \left(\sqrt{2}\right)^2$$

is a **difference of two squares**.

3 $2x^2 - 3x + 1 = 2\left(x^2 - \dfrac{3}{2}x\right) + 1$

$$= 2\left(x^2 - \frac{3}{2}x + \left(\frac{3}{4}\right)^2\right) - 2 \times \left(\frac{3}{4}\right)^2 + 1$$

$$= 2\left(x - \frac{3}{4}\right)^2 - \frac{1}{8} = \left\{\sqrt{2}\left(x - \frac{3}{4}\right)\right\}^2 - \left(\sqrt{\frac{1}{8}}\right)^2$$

is a **difference of two squares**.

4 $x^2 + 6x + 14 = (x^2 + 6x + (3)^2) - 3^2 + 14$

$$= (x + 3)^2 + 5 = (x + 3)^2 + \left(\sqrt{5}\right)^2$$

is a **sum of two squares**.

5 $12 + 5x - x^2 = 12 - (x^2 - 5x)$

$$= 12 - \left\{x^2 - 5x + \left(\frac{5}{2}\right)^2\right\} + \frac{25}{4}$$

$$= \frac{73}{4} - \left(x - \frac{5}{2}\right)^2 = \left(\frac{\sqrt{73}}{2}\right)^2 - \left(x - \frac{5}{2}\right)^2$$

is a **difference of two squares**.

These situations with different denominators are considered separately in the next sections.

13.3 Denominator: a perfect square

When a fraction has a denominator of the form $(x + a)^2$, you can integrate it if the numerator is a constant. Use the method of partial fractions (which you will find in an algebra book), to rewrite it in the form in which the numerator is a constant. Then you can integrate it.

Example 13.3.1

Evaluate $\int \dfrac{3x+1}{(x+1)^2} dx$

Use partial fractions to rewrite $\dfrac{3x+1}{(x+1)^2}$ as $\dfrac{3}{x+1} - \dfrac{2}{(x+1)^2}$

Then
$$\int \frac{3x+1}{(x+1)^2} dx = \int \left(\frac{3}{x+1} - \frac{2}{(x+1)^2} \right) dx$$
$$= 3\ln|x+1| + \frac{2}{x+1} + C$$

where you write down the second integral using $\int \dfrac{dx}{x^2} = -\dfrac{1}{x}$

13.4 Denominator: a difference of squares

When the denominator is the difference of two squares you can always write it as a product of factors, even if sometimes the factors are not very pleasant. For example, using the cases of the difference of two squares in Example 13.2.1 you find:

i $x^2 + 4x + 2 = (x+2)^2 - (\sqrt{2})^2$
$$= (x+2-\sqrt{2})(x+2+\sqrt{2})$$

ii $2x^2 - 3x + 1 = (2x-1)(x-1)$

iii $12 + 5x - x^2 = \left(\dfrac{\sqrt{73}}{2} \right)^2 - \left(x - \dfrac{5}{2} \right)^2$

$$= \left(\frac{\sqrt{73}}{2} - \left(x - \frac{5}{2} \right) \right) \left(\frac{\sqrt{73}}{2} + \left(x - \frac{5}{2} \right) \right)$$

$$= \left(\frac{\sqrt{73} + 5}{2} - x \right) \left(\frac{\sqrt{73} - 5}{2} + x \right)$$

Once you have a product of factors, you should split the expression into its partial fractions (using an algebra book for reference if necessary).

Nugget – Ah! it's my old friend, the difference of two squares!

As you may already be aware, the phrase 'the difference of two squares' refers to the fact that an expression in this form, say, $a^2 - b^2$ can be factorized into $(a + b)(a - b)$. For example, $9x^2 - 16$ can be factorized into $(3x + 4)(3x - 4)$, and so on. It is remarkable how often an awareness of this 'trick' is the key to unlocking the solution to mathematics questions. Keep this in mind as you work through the worked example below and also when you tackle Exercise 13.2.

Example 13.4.1

Integrate $\int \frac{x + 35}{x^2 - 25} dx$

Using partial fractions, you can rewrite $\frac{x + 35}{x^2 - 25}$ as $\frac{4}{x - 5} - \frac{3}{x + 5}$

$$\int \frac{x + 35}{x^2 - 25} dx = \int \left(\frac{4}{x - 5} - \frac{3}{x + 5} \right) dx$$

$$= 4 \ln|x - 5| - 3 \ln|x + 5| + C$$

Example 13.4.2

Integrate $\int \frac{dx}{x^2 - a^2}$

Using partial fractions, rewrite $\frac{1}{x^2 - a^2}$ as $\frac{1}{2a} \times \frac{1}{x - a} - \frac{1}{2a} \times \frac{1}{x + a}$

$$\int \frac{dx}{x^2 - a^2} = \int \left(\frac{1}{2a} \times \frac{1}{x-a} - \frac{1}{2a} \times \frac{1}{x+a} \right) dx$$

$$= \frac{1}{2a} \ln|x-a| - \frac{1}{2a} \ln|x+a| + C$$

Example 13.4.3

Integrate $\int \frac{23 - 2x}{2x^2 + 9x - 5} dx$

Rewrite $\dfrac{23 - 2x}{2x^2 + 9x - 5}$ as $\dfrac{4}{2x-1} - \dfrac{3}{x+5}$

$$\int \frac{23 - 2x}{2x^2 + 9x - 5} dx = \int \left(\frac{4}{2x-1} - \frac{3}{x+5} \right) dx$$

$$= 2 \ln|2x-1| - 3 \ln|x+5| + C$$

Example 13.4.4

Evaluate $\int \frac{x^2 + 10x + 6}{x^2 + 2x - 8} dx$

This time you need to divide first as in Examples 13.1.1 and 13.1.2.

$$\frac{x^2 + 10x + 6}{x^2 + 2x - 8} = 1 + \frac{8x + 14}{x^2 + 2x - 8}$$

Now you can use partial fractions.

$$\frac{x^2 + 10x + 6}{x^3 + 2x - 8} = 1 + \frac{5}{x-2} + \frac{3}{x+4}$$

$$\int \frac{x^2 + 10x + 6}{x^2 + 2x - 8} dx = \int \left(1 + \frac{5}{x-2} + \frac{3}{x+4} \right) dx$$

$$= x + 5 \ln|x-2| + 3 \ln|x+4| + C$$

In the examples in the previous sections of this chapter, you have seen that you should rewrite rational functions in partial fraction form, dividing first if necessary. However, sometimes it isn't possible to factorize the quadratic factor.

Exercise 13.2

Find the following integrals.

1 $\displaystyle\int \frac{dx}{x^2-1}$

2 $\displaystyle\int \frac{dx}{1-x^2}$

3 $\displaystyle\int \frac{x^2 dx}{x^2-4}$

4 $\displaystyle\int \frac{dx}{4x^2-9}$

5 $\displaystyle\int \frac{x+8}{x^2+6x+8}dx$

6 $\displaystyle\int \frac{3x-1}{x^2+x-6}dx$

7 $\displaystyle\int \frac{8x+1}{2x^2-9x-35}dx$

8 $\displaystyle\int \frac{x+1}{3x^2-x-2}dx$

9 $\displaystyle\int \frac{7x-8}{4x^2+3x-1}dx$

10 $\displaystyle\int \frac{1+x}{(1-x)^2}dx$

11 $\displaystyle\int \frac{2x-1}{(x+2)^2}dx$

12 $\displaystyle\int \frac{2x+1}{(2x+3)^2}dx$

13 $\displaystyle\int \frac{x^2-2}{x^2-x-12}dx$

14 $\displaystyle\int \frac{x^2+1}{x^2-x-2}dx$

13.5 Denominator: a sum of squares

When the denominator of the algebraic fraction is the sum of two squares, you cannot factorize it, and you need to use a different strategy for integration.

The method is to use two standard forms:

$$\int \frac{dx}{a^2+x^2} = \frac{1}{a}\tan^{-1}\frac{x}{a}+C \quad \text{and} \quad \int \frac{x\,dx}{a^2+x^2} = \frac{1}{2}\ln(a^2+x^2)+C$$

To demonstrate the method, here are some worked examples. In some of them you need first to do some preliminary work before you can use the standard forms to carry out the integration.

13. *Integration of algebraic fractions* (215)

Example 13.5.1

Integrate $\int \dfrac{(2x+3)}{4+x^2}dx$

First split the numerator into two parts, remembering that:

$$\int \frac{2x}{4+x^2}dx = \ln(x^2+4) \text{ and } \int \frac{1}{4+x^2}dx = \frac{1}{2}\tan^{-1}\frac{x}{2}$$

Thus

$$\int \frac{(2x+3)}{4+x^2}dx = \int \frac{2x}{4+x^2}dx + \int \frac{3}{4+x^2}dx$$

$$= \int \frac{2x}{4+x^2}dx + 3\int \frac{1}{4+x^2}dx$$

$$= \ln(x^2+4) + \frac{3}{2}\tan^{-1}\frac{x}{2} + C$$

Example 13.5.2

Integrate $\int \dfrac{(3x+2)dx}{x^2+6x+14}$

First note that the denominator can be written as a sum of two squares in the form $x^2+6x+14 = (x+3)^2 + (\sqrt{5})^2$

Then make the substitution $x+3 = z$, so that the denominator becomes z^2+5. Then, as $dx = dz$, you find that:

$$\int \frac{(3x+2)dx}{x^2+6x+14} = \int \frac{3(z-3)+2}{z^2+5}dz$$

$$= \int \frac{3z-7}{z^2+5}dz$$

$$= \frac{3}{2}\int \frac{2z}{z^2+5}dz - 7\int \frac{1}{z^2+5}dz$$

$$= \frac{3}{2}\ln(z^2+5) - \frac{7}{\sqrt{5}}\tan^{-1}\frac{z}{\sqrt{5}} + C$$

$$= \frac{3}{2}\ln(x^2+6x+14) - \frac{7}{\sqrt{5}}\tan^{-1}\frac{x+3}{\sqrt{5}} + C$$

Example 13.5.3

Integrate $\int \dfrac{(5x+1)dx}{2x^2+4x+3}$

First note that the denominator can be written as a sum of two squares in the form $2x^2 + 4x + 3 = 2(x^2 + 2x + 1) + 1 = (\sqrt{2}(x+1))^2 + 1$

Then make the substitution $\sqrt{2}(x+1) = z$, so that the denominator becomes $z^2 + 1$. Then, as $x = \dfrac{z}{\sqrt{2}} - 1$ and $dx = \dfrac{1}{\sqrt{2}}\,dz$, you find:

$$\int \frac{(5x+1)dx}{2x^2+4x+3} = \int \frac{5\left(\dfrac{z}{\sqrt{2}}-1\right)+1}{z^2+1}\frac{1}{\sqrt{2}}\,dz$$

$$= \frac{5}{2}\int \frac{z}{z^2+1}dz - \int \frac{4}{z^2+1}\frac{1}{\sqrt{2}}dz$$

$$= \frac{5}{4}\int \frac{2z}{z^2+1}dz - \frac{4}{\sqrt{2}}\int \frac{1}{z^2+1}dz$$

$$= \frac{5}{4}\ln(z^2+1) - \frac{4}{\sqrt{2}}\tan^{-1}z + C$$

$$= \frac{5}{4}\ln(2x^2+4x+3) - 2\sqrt{2}\tan^{-1}(\sqrt{2}(x+1)) + C$$

13.6 Denominators of higher degree

If the denominator is a cubic function of x, then use the method of partial fractions to reduce their degrees. You will then be able to use the methods of earlier sections.

Example 13.6.1

Integrate $\int \dfrac{(3-4x-x^2)}{x(x^2-4x+3)}dx$

Using partial fractions

$$\frac{(3-4x-x^2)}{x(x^2-4x+3)} = \frac{1}{x} + \frac{1}{x-1} - \frac{3}{x-3}$$

Then
$$\int \frac{(3-4x-x^2)}{x(x^2-4x+3)}dx = \int \frac{1}{x}dx + \int \frac{1}{x-1}dx - \int \frac{3}{x-3}dx$$
$$= \ln|x| + \ln|x-1| - 3\ln|x-3| + C$$

Example 13.6.2

Integrate $\int \frac{(x-1)}{(x+1)(x^2+1)}dx$

Using partial fractions

$$\frac{(x-1)}{(x+1)(x^2+1)} = -\frac{1}{x+1} + \frac{x}{x^2+1}$$

Then
$$\int \frac{(x-1)}{(x+1)(x^2+1)}dx = \int -\frac{1}{x+1}dx + \int \frac{x}{x^2+1}dx$$
$$= -\ln|x+1| + \tfrac{1}{2}\ln|x^2+1| + C$$

Exercise 13.3

Carry out the following integrations.

1 $\int \dfrac{1}{x^2+6x+17}dx$ **2** $\int \dfrac{1}{x^2+6x-4}dx$

3 $\int \dfrac{1}{x^2+4x+6}dx$ **4** $\int \dfrac{1}{2x^2+2x+7}dx$

5 $\int \dfrac{1-3x}{3x^2+4x+2}dx$ **6** $\int \dfrac{(4x-5)dx}{x^2-2x-1}$

7 $\int \dfrac{(2x+5)dx}{x^2+4x+5}$ **8** $\int \dfrac{dx}{x^3+1}$

9 $\int \dfrac{(x-1)^2dx}{x^2+2x+2}$ **10** $\int \dfrac{(3x+5)dx}{1-2x-x^2}$

13.7 Denominators with square roots

This section covers integrals of the form:

$$\int \frac{(Ax + B)}{\sqrt{px^2 + qx + r}}\, dx$$

Using the same type of method as earlier in the chapter, you can reduce this form to one of the following forms:

Case 1a $\int \dfrac{z}{\sqrt{z^2 + a^2}}\, dz$ **Case 1b** $\int \dfrac{1}{\sqrt{z^2 + a^2}}\, dz$

Case 2a $\int \dfrac{z}{\sqrt{z^2 - a^2}}\, dz$ **Case 2b** $\int \dfrac{1}{\sqrt{z^2 - a^2}}\, dz$

Case 3a $\int \dfrac{z}{\sqrt{a^2 - z^2}}\, dz$ **Case 3b** $\int \dfrac{1}{\sqrt{a^2 - z^2}}\, dz$

These cases are now dealt with in turn.

Case 1a
$$\int \frac{z}{\sqrt{z^2 + a^2}}\, dz$$

This is an example of a function of $z^2 + a^2$, and the integral is:

$$\int \frac{z}{\sqrt{z^2 + a^2}}\, dz = (z^2 + a^2)^{\frac{1}{2}} + C$$

Case 1b
$$\int \frac{1}{\sqrt{z^2 + a^2}}\, dz$$

Let $z = a \tan \theta$. Then $\sqrt{z^2 + a^2} = a \sec \theta$ and $dz = a \sec^2 \theta\, d\theta$

$$\int \frac{z}{\sqrt{z^2 + a^2}}\, dz = \int \frac{1}{a \sec \theta} \times a \sec^2 \theta\, d\theta$$

$$= \int \sec \theta\, d\theta$$

$$= \ln|\sec \theta + \tan \theta| + C$$

$$= \ln\left| \frac{\sqrt{z^2 + a^2}}{a} + \frac{z}{a} \right| + C$$

Case 2a
$$\int \frac{z}{\sqrt{z^2 - a^2}}\, dz$$

As in Case 1a above,

$$\int \frac{z}{\sqrt{z^2 - a^2}}\, dz = (z^2 - a^2)^{\frac{1}{2}} + C$$

Case 2b $$\int \frac{1}{\sqrt{z^2 - a^2}}\, dz$$

Let $= a \sec\theta$. Then $\sqrt{z^2 - a^2} = a \tan\theta$ and $dz = a \sec\theta \tan\theta\, d\theta$.

Then $$\int \frac{1}{\sqrt{z^2 + a^2}}\, dz = \int \frac{1}{a \tan\theta} \times a \sec\theta \tan\theta\, d\theta$$

$$= \int \sec\theta\, d\theta$$

$$= \ln|\sec\theta + \tan\theta| + C$$

$$= \ln\left|\frac{z}{a} + \frac{\sqrt{z^2 - a^2}}{a}\right| + C$$

Case 3a $$\int \frac{z}{\sqrt{a^2 - z^2}}\, dz$$

As in Cases 1a and 2a above,

$$\int \frac{z}{\sqrt{a^2 - z^2}}\, dz = -(a^2 - z^2)^{\frac{1}{2}} + C$$

Case 3b $$\int \frac{1}{\sqrt{a^2 - z^2}}\, dz$$

This is a standard form, see Section 11.10, number 17.

$$\int \frac{1}{\sqrt{a^2 - z^2}}\, dz = \sin^{-1}\frac{z}{a} + C$$

You should recognize Cases 1a, 2a and 3a, and you should remember Case 3b. It is easiest to work out the others as you meet them.

Example 13.7.1

Integrate $\displaystyle\int \frac{dx}{\sqrt{x^2+6x+10}}$

Writing $x^2 + 6x + 10 = (x+3)^2 + 1$ and letting $z = x + 3$ you find that:

$$\int \frac{dx}{\sqrt{x^2+6x+10}} = \int \frac{dz}{\sqrt{z^2+1}}$$

$$= \ln\left|\sqrt{z^2+1}+z\right| + C$$

$$= \ln\left|\sqrt{x^2+6x+10}+x+3\right| + C$$

Example 13.7.2

Integrate $\displaystyle\int \frac{x+1}{\sqrt{2x^2+x-3}}\,dx$

Write $2x^2 + x - 3 = 2\left(x+\frac{1}{4}\right)^2 - \frac{25}{8}$ and letting

$\sqrt{2}\left(x+\frac{1}{4}\right) = \frac{5}{2\sqrt{2}}z$ or $x = \frac{5}{4}z - \frac{1}{4}$ you find that:

$$\int \frac{x+1}{\sqrt{2x^2+x-3}}\,dx = \int \frac{\left(\left(\frac{5}{4}z-\frac{1}{4}\right)+1\right)}{\sqrt{\frac{25}{8}z^2-\frac{25}{8}}}\,\frac{5}{4}\,dz$$

$$= \frac{5}{4}\times\frac{5}{4}\times\sqrt{\frac{8}{25}}\int \frac{z\,dz}{\sqrt{z^2-1}} \times \frac{3}{4}\times\frac{5}{4}\times\sqrt{\frac{8}{25}}\int \frac{dz}{\sqrt{z^2-1}}$$

$$= \frac{5}{4\sqrt{2}}\sqrt{z^2-1}+\frac{3}{4\sqrt{2}}\ln\left|z+\sqrt{z^2-1}\right| + C$$

After re-substituting for z and some simplification you find that:

$$\int \frac{x+1}{\sqrt{2x^2+x-3}}\,dx$$

$$= \frac{1}{2}\sqrt{2x^2+x-3}+\frac{3}{4\sqrt{2}}\ln\left|\frac{4x+1}{5}+\sqrt{\frac{8(2x^2+x-3)}{25}}\right| + C$$

There are alternative forms for this answer.

Exercise 13.4

Carry out the following integrations.

1 $\int \dfrac{dx}{\sqrt{x^2 + 6x + 10}}$ 　　　　**2** $\int \dfrac{dx}{\sqrt{x^2 + 2x + 4}}$

3 $\int \dfrac{dx}{\sqrt{x^2 - 4x + 2}}$ 　　　　**4** $\int \dfrac{dx}{\sqrt{1 - x - x^2}}$

5 $\int \dfrac{dx}{\sqrt{x(4 - x)}}$ 　　　　**6** $\int \dfrac{x \, dx}{\sqrt{x^2 + 1}}$

7 $\int \dfrac{(x + 1)dx}{\sqrt{x^2 - 1}}$ 　　　　**8** $\int \dfrac{(x + 1)dx}{\sqrt{x^2 - 1}}$

9 $\int \dfrac{x \, dx}{\sqrt{x^2 - x + 1}}$ 　　　　**10** $\int \dfrac{(2x - 3)dx}{\sqrt{x^2 - 2x + 5}}$

11 $\int \dfrac{(2x + 1)dx}{\sqrt{3 - 4x - x^2}}$ 　　　　**12** $\int \dfrac{(x + 2)dx}{\sqrt{x^2 + 2x - 1}}$

Key ideas

▶ In this chapter, techniques for integrating certain types of algebraic fractions were offered. The first of these was *rational fractions* (Section 13.1).

▶ Integrating an algebraic fraction where the denominator is of the second degree or higher is awkward. The method of *partial fractions* involves rearranging the fraction so that the denominator can be resolved into linear fractions that are different (Sections 13.3 and 13.4).

▶ Section 13.6 show how to deal with cases where the denominator contains a quadratic factor that cannot itself be factorized.

▶ Section 13.7 show how to deal with cases where the denominator is irrational (i.e. it is contained within a square root).

14

Area and definite integrals

In this chapter you will learn:
- ▶ *how to use integration to find area*
- ▶ *what a definite integral is*
- ▶ *under what circumstances integration is possible when the function approaches infinity.*

14.1 Areas by integration

The integral calculus has its origin in the attempt to find a general method for calculating areas of shapes. When these shapes are bounded by straight lines, you can use geometrical methods to obtain formulae for their areas; but when the boundaries are wholly or partly curves (such as the circle or parabola) then, unless you use the integral calculus, you have to depend upon experimental or approximate methods.

Consider as an example, the parabola $y = x^2$

Figure 14.1 shows part of this parabola. Let C be any point on the curve and draw CD parallel to the y-axis to meet the x-axis at D.

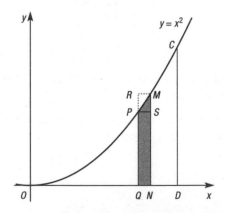

Figure 14.1

Let $OD = a$ units

Suppose that you need to find the area under OC, that is, the area of OCD, which is bounded by the curve, the x-axis and CD.

Let P be any point on the curve and let the coordinates of P be (x, y).

If you draw the line PQ parallel to the y-axis to meet the x-axis at Q you have $OQ = x$, $PQ = y$. Now let M be another point on the curve, close to P, such that the x-coordinate of M is $x + \delta x$. Draw MN parallel to the y-axis to meet the x-axis at N.

Draw PS and MR parallel to the x-axis and produce QP so that it meets MR at R. Then $MS = \delta y$.

Suppose that the shaded area bounded by PQ, QN, NM and the piece of curve MP is denoted by δA.

The area δA lies between the rectangular areas $QPSN$ and $QRMN$, which are $y\delta x$ and $(y + \delta y)\delta x$ respectively.

Therefore
$$y\delta x < \delta A < (y + \delta y)\delta x$$

and
$$y < \frac{\delta A}{\delta x} < y + \delta y$$

Now suppose δx is decreased indefinitely.

Then as $\delta x \to 0$, $\delta y \to 0$, and $\dfrac{\delta A}{\delta x} \to \dfrac{dA}{dx}$

Therefore
$$\frac{dA}{dx} - y$$

$$= x^2$$

and
$$A = \tfrac{1}{3}x^3 + C$$

This result provides a formula for the area A in terms of x and the undetermined constant C.

But, if you measure the area A from the origin O, when $x = 0$, $A = 0$. Substituting these values in the equation for A gives:

$$0 = \tfrac{1}{3}0^3 + C$$

so
$$C = 0$$

Then
$$A = \tfrac{1}{3}x^3$$

When $x = a$ as in Figure 14.2 for the area of OCD, then

$$A = \tfrac{1}{3}a^3$$

If you take another value of x, say $x = b$, so that $OE = b$ in Figure 14.2, then the area of OEF is $\tfrac{1}{3}b^3$.

Therefore the area of the shaded region $CDEF$ is $\tfrac{1}{3}a^3 - \tfrac{1}{3}b^3$

In Section 14.2, a general rule which applies to finding the area under any function will be established.

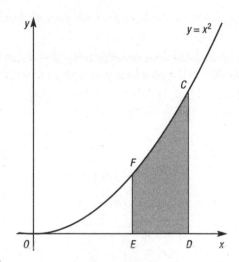

Figure 14.2

14.2 Definite integrals

Nugget – The overall strategy for finding areas under curves

As you will see in this section, the general strategy for finding the area under a curve can be summarized roughly as follows (the argument is actually slightly more subtle than this, but here is the gist of it – keep Figure 14.3 handy as you read).

Suppose the curve drawn in Figure 14.3 is part of the graph of
$y = \phi(x)$.

Let *CD* and *FE* be fixed lines parallel to the *y*-axis such that
$OD = b$ and $OE = a$. Suppose that you wish to find the area of
the shaded region *FEDC*. Let this area be *A*.

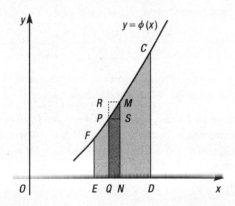

Figure 14.3

Let *P* be any point on the curve and let the coordinates of *P*
be (x, y). Drawing the line *PQ* parallel to the *y*-axis to meet the
x-axis at *Q* you have $OQ = x$, $PQ = y = \phi(x)$. Now let *M* be
another point on the curve, close to *P*. Draw *MN* parallel to
the *y*-axis to meet the *x*-axis at *N*. Draw *PS* and *MR* parallel
to the *x*-axis and produce *QP* so that it meets *MR* at *R*.

Then $MS = \delta y$

                        $MN = y + \delta y$

and $ON = x + \delta x$

Suppose that the shaded area bounded by PQ, QN, NM and the piece of curve MP is denoted by δA.

The area δA lies between the rectangular areas $QPSN$ and $QRMN$, which are $y\delta x$ and $(y + \delta y)\delta x$ respectively.

Therefore $\qquad y\delta x < \delta A < (y + \delta y)\delta x$

and $\qquad\qquad\qquad y < \dfrac{\delta A}{\delta x} < y + \delta y$

Now suppose δx is decreased indefinitely.

Then as $\delta x \to 0$, $\delta y \to 0$, and $\frac{\delta A}{\delta x} \to \frac{dA}{dx}$

Therefore $\qquad\qquad\qquad\qquad \dfrac{dA}{dx} = y$

$$= \phi(x)$$

Integrating, and representing the integral of $\phi(x)$ by $f(x)$, you find:

$$A = \int \phi(x)\,dx$$
$$= f(x) + C$$

where C is an undetermined constant. You can find its value if you know the value of A for some value of x.

Now A is the area $FEDC$, that is, between the lines where $x = a$ and $x = b$ respectively. But when $x = a$, $A = 0$ substituting this in the equation $A = f(x) + C$ you find that $0 = f(a) + C$, so that $C = -f(a)$. When $x = b$, that is, at D

$$A = f(b) + C$$

Substituting the value found for C

$$A = f(b) - f(a)$$

You can find $f(a)$ and $f(b)$ by substituting a and b for x into the equation for $f(x)$ which is the integral of $\phi(x)$.

The area, A, between the limits a and b can be found by integrating $\phi(x)$ and substituting the values $x = a$ and $x = b$, and then subtracting $f(a)$ from $f(b)$. This can conveniently be expressed by the notation

$$f(b) - f(a) = \int_a^b \phi(x)\,dx$$

$\int_a^b \phi(x)\,dx$ is called a **definite integral** and a and b are called its limits, a being the **lower limit** and b the **upper limit**.

To evaluate a definite integral such as $\int_a^b \phi(x)\,dx$

▶ find the indefinite integral $\int \phi(x)\,dx$, that is, $f(x)$

▶ find $f(b)$ by substituting $x = b$ in $f(x)$

▶ find $f(a)$ by substituting $x = a$ in $f(x)$

▶ subtract $f(a)$ from $f(b)$.

In practice the following notation and arrangement is used:

$$\int_a^b \phi(x)\,dx = [f(x)]_a^b$$
$$= f(b) - f(a)$$

14.3 Characteristics of a definite integral

Note the following points about a definite integral.

▶ The results of substituting the limits a and b in the integral are respectively $f(a) + C$ and $f(b) + C$. Consequently on subtraction the constant C disappears; this is the reason why the integral is called 'definite'.

▶ The variable is assumed to be increasing from the lower limit to the upper limit, that is, in the above from a to b. You must remember this when you deal with negative limits. If, for example, the limits are -2 and 0, then the variable x is increasing from -2 to 0. Consequently the upper limit is 0 and the lower limit -2.

▶ This definite integral is therefore written as

$$\int_{-2}^{0} \phi(x)\,dx$$

▶ The term 'limit' in this connection is different from the meaning attached to it in Section 2.2. It denotes the values of x at the ends of the interval of values over which you are finding the value of the definite integral.

Nugget – ∫ is for SUM

If you are able to exploit the mental image of integration being the sum of a series of very thin rectangular areas bounded by the curve and the x-axis, then this helps to make more sense of why the sign for an integral, $\int$, can be thought of as a giant letter S, where S stands for 'sum'. The notation was first used by the German mathematician, Leibniz, towards the end of the 17th century.

Example 14.3.1

Evaluate the definite integral $\int_2^5 3x\,dx$

Now
$$\int 3x\,dx = \frac{3}{2}x^2 + C$$

Therefore
$$\int_2^5 3x\,dx = \left[\frac{3}{2}x^2\right]_2^5$$

$$= \frac{3}{2}(5^2 - 2^2)$$

$$= \frac{3}{2} \times 21$$

$$= \frac{63}{2} = 31\frac{1}{2}$$

You will find it useful to check this by drawing the graph of $y = 3x$, the lines $x = 2$ and $x = 5$ and finding the area of the trapezium by using the ordinary geometrical rule.

Example 14.3.2

Evaluate the definite integral $\int_0^{\frac{1}{2}\pi} \sin x\,dx$

Now
$$\int \sin x\,dx = -\cos x + C$$

Therefore
$$\int_0^{\frac{1}{2}\pi} \sin x \, dx = [-\cos x]_0^{\frac{1}{2}\pi}$$

$$= \left\{\left(-\cos\frac{1}{2}\pi\right) - (-\cos 0)\right\}$$

$$= 0 + 1$$

$$= 1$$

Notice that this gives, in square units, the area beneath the curve $y = \sin x$ between 0 and $\frac{1}{2}\pi$. Figure 14.4 shows the graph of this function between $x = 0$ and $x = \pi$.

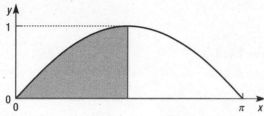

Figure 14.4

From symmetry, the area under this curve between 0 and π must be twice that between 0 and $\frac{1}{2}\pi$. that is, 2 square units.

You can check this by calculating $\int_0^\pi \sin x \, dx$

Example 14.3.3

Calculate $\int_0^1 xe^{x^2} \, dx$

By inspection
$$\int xe^{x^2} \, dx = \frac{1}{2}e^{x^2} + C$$

Therefore
$$\int_0^1 xe^{x^2} \, dx = \left[\frac{1}{2}e^{x^2}\right]_0^1$$

$$= \frac{1}{2}\left(e^1 - e^0\right)$$

$$= \frac{1}{2}(e - 1)$$

Example 14.3.4

Evaluate the definite integral $\int_{-1}^{0}\left(1+3x-2x^2\right)dx$

$$\int\left(1+3x-2x^2\right)dx = x+\frac{3}{2}x^2-\frac{2}{3}x^3+C$$

so

$$\int_{-1}^{0}\left(1+3x-2x^2\right)dx = \left[x+\frac{3}{2}x^2-\frac{2}{3}x^3\right]_{-1}^{0}$$

$$=0-\left(-1+\frac{3}{2}+\frac{2}{3}\right)$$

$$=-\frac{7}{6}$$

Example 14.3.5

Evaluate the definite integral $\int_{2}^{3}\dfrac{dx}{\sqrt{x^2-1}}$

so

$$\int\frac{dx}{\sqrt{x^2-1}} = \cosh^{-1}x+C$$

$$\int_{2}^{3}\frac{dx}{\sqrt{x^2-1}} = \left[\cosh^{-1}x\right]_{2}^{3}$$

$$=\cosh^{-1}3-\cosh^{-1}2$$

$$=0.446 \text{ (approximately)}$$

Example 14.3.6

Evaluate the definite integral $\int_{1}^{e}x\ln x\, dx$

$$\int x\ln x\, dx = \frac{1}{2}x^2\ln x-\frac{x^2}{4}+C$$

so

$$\int_{1}^{e}x\ln x\, dx = \left[\frac{1}{2}x^2\ln x-\frac{1}{4}x^2\right]_{1}^{e}$$

$$=\left(\frac{1}{2}e^2\ln e-\frac{1}{4}e^2\right)-\left(\frac{1}{2}1^2\ln 1-\frac{1}{4}1^2\right)$$

$$=\frac{1}{2}e^2-\frac{1}{4}e^2+\frac{1}{4}=\frac{1}{4}e^2+\frac{1}{4}$$

Exercise 14.1

Evaluate the following integrals.

1 $\int_1^4 x^2\,dx$

2 $\int_0^1 (x^2 + 4)\,dx$

3 $\int_1^2 (x^2 + 3x - 5)\,dx$

4 $\int_{-2}^1 (2x + 1)^2\,dx$

5 $\int_1^{10} x^{-0.8}\,dx$

6 $\int_1^4 \sqrt{x}\,dx$

7 $\int_0^4 \left(x^{\frac{1}{2}} + x^{-\frac{1}{2}}\right)dx$

8 $\int_0^{\frac{1}{6}\pi} \cos 3x\,dx$

9 $\int_0^{\frac{1}{2}\pi} (\cos\theta - \sin 2\theta)\,d\theta$

10 $\int_0^{\frac{1}{4}\pi} \cos\left(2\theta + \frac{1}{4}\pi\right)d\theta$

11 $\int_{-1}^1 2^x\,dx$

12 $\int_0^2 e^{\frac{1}{2}x}\,dx$

13 $\int_0^{\frac{1}{2}\pi} \frac{1}{2}r^2\,d\theta$

14 $\frac{1}{2}\pi\rho\int_{-a}^a \left(a^2 - x^2\right)^2 dx$

15 $\int_a^b e^{kx}\,dx$

16 $\int_0^{\frac{1}{2}\pi} \sin^2 x\,dx$

17 $\int_2^3 \frac{x}{1 + x^2}\,dx$

18 $\int_0^{\frac{1}{2}\pi} x\sin x\,dx$

19 $\int_0^1 x\ln x\,dx$

20 $\int_0^1 x^2 \ln x\,dx$

14.4 Some properties of definite integrals

1 $\int_a^b f(x)\,dx = -\int_b^a f(x)\,dx$

Let $\phi(x)$ be the indefinite integral of $f(x)$. Then, if the limits of the definite integral are a and b $\int_a^b f(x)\,dx = \phi(b) - \phi(a)$

If you interchange the limits $\int_b^a f(x)\,dx = \phi(a) - \phi(b)$

that is
$$\int_a^b f(x)\,dx = -\int_b^a f(x)\,dx$$

Thus, interchanging the limits of integration changes the sign of the definite integral.

2 $\int_a^b f(x)\,dx = \int_a^c f(x)\,dx + \int_c^b f(x)\,dx$

Let $\phi(x)$ be the indefinite integral of $f(x)$.

Then
$$\int_a^b f(x)\,dx = \phi(b) - \phi(a)$$

and $\int_a^c f(x)\,dx = \phi(c) - \phi(a)$ and $\int_c^b f(x)\,dx = \phi(b) - \phi(c)$

Therefore $\int_a^c f(x)\,dx + \int_c^b f(x)\,dx = (\phi(c) - \phi(a)) + (\phi(b) - \phi(c))$

$$= \phi(b) - \phi(a)$$
$$= \int_a^b f(x)\,dx$$

Look back to Figure 14.2 which illustrates this theorem. In Figure 14.2:

area of OCD = area of OEF + area of $FEDC$

that is $\int_0^a f(x)\,dx = \int_0^b f(x)\,dx + \int_b^a f(x)\,dx$

3 Notice that $\int_a^b f(x)\,dx = \phi(b) - \phi(a)$ where $\phi(x)$ is the indefinite integral of $f(x)$, does not contain x. Therefore any other letter could be used in the integral, provided the function of each of the two letters in the sum is the same.

For the function f $\int_a^b f(y)\,dy = \phi(b) - \phi(a)$

and $\int_a^b f(x)\,dx = \phi(b) - \phi(a)$

Therefore $\int_a^b f(x)\,dx = \int_a^b f(y)\,dy$

4 $\displaystyle\int_0^a f(x)\,dx = \int_0^a f(a-x)\,dx$

Let $x = a - u$ or $a - x = u$. Then $dx = -\,du$.

Now, if in definite integration the variable is changed, the limits will also be changed and the new limits must be determined.

Therefore, in the integral $\displaystyle\int_a^b f(x)\,dx$, when $x = 0$, $u = a - x = a - 0 = a$ and when $x = a$, $u = a - x = a - a = 0$. Thus when $x = 0$, $u = a$ and when $x = a$, $u = 0$.

Therefore when x is replaced by $a - u$ in $\displaystyle\int_0^a f(x)\,dx$, the limits must be changed from 0 and a to a and 0.

Therefore
$$\int_0^a f(x)\,dx = -\int_a^0 f(a-u)\,du$$

$$= \int_0^a f(a-u)\,du \qquad \text{by 1 above}$$

$$= \int_0^a f(a-x)\,dx \qquad \text{by 3 above}$$

Example 14.4.1

$$\int_0^{\frac{1}{2}\pi} \sin x\,dx = \int_0^{\frac{1}{2}\pi} \sin\!\left(\frac{1}{2}\pi - x\right) dx = \int_0^{\frac{1}{2}\pi} \cos x\,dx$$

In general
$$\int_0^{\frac{1}{2}\pi} f(\sin x)\,dx = \int_0^{\frac{1}{2}\pi} f(\cos x)\,dx$$

14.5 Infinite limits and infinite integrals

In calculating definite integrals between two limits a and b it has been assumed:

▶ that these limits are finite

▶ that all the values of the function between them are also finite, that is, the function is continuous.

However, there are cases when one or both of these conditions is not satisfied.

14.6 Infinite limits

The problems that arise when one of the limits is infinite are illustrated by the case $y = \frac{1}{x^2}$. It is useful to look at the graph of $y = \frac{1}{x^2}$ shown in Figure 14.5.

As the values of $\frac{1}{x^2}$ are always positive, the graph of the function lies entirely above the x-axis. It consists of two parts, corresponding to positive and negative values of x. These two parts are symmetrical about the y-axis.

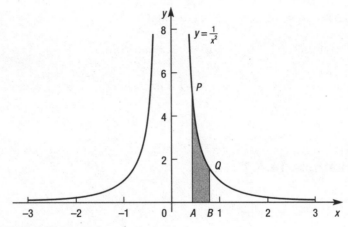

Figure 14.5 Graph of $y = \frac{1}{x^2}$

Let P and Q be two points on the curve, and let PA and QB be the corresponding ordinates. Let $OA = a$ and $OB = b$. Then, as shown in Section 14.2, the area of the region beneath the part of the curve PQ and bounded by PA, QB and the x-axis, shown by the shaded part of Figure 14.5, is represented by

$$\int_a^b \frac{dx}{x^2} = \left[-\frac{1}{x} \right]_a^b = -\left(\frac{1}{b} - \frac{1}{a} \right)$$

1   Suppose the ordinate QB moves indefinitely away from the y-axis, so that OB, that is b, increases indefinitely. Then the ordinate QB decreases indefinitely and in the limit the x-axis is an asymptote to the curve (see Section 2.1); that is, as $b \to \infty$, $QB \to 0$.

As $b \to \infty$ the integral becomes:

$$\lim_{b\to\infty} \int_a^b \frac{dx}{x^2} = \lim_{b\to\infty}\left\{-\left(\frac{1}{b}-\frac{1}{a}\right)\right\} = \frac{1}{a}$$

Therefore the integral is finite and takes the value $\frac{1}{a}$.

The notation $\int_a^\infty \frac{dx}{x^2}$ is used as a shorthand for $\lim_{b\to\infty} \int_a^b \frac{dx}{x^2}$

Then
$$\int_a^\infty \frac{dx}{x^2} = \lim_{b\to\infty} \int_a^b \frac{dx}{x^2}$$

$$= \lim_{b\to\infty}\left\{-\left(\frac{1}{b}-\frac{1}{a}\right)\right\} = \frac{1}{a}$$

2   Next, suppose that the ordinate PA moves towards the y-axis; then PA increases rapidly, and when OA, that is a, decreases indefinitely the corresponding value of y increases without limit.

The definite integral can now be written:

$$\lim_{a\to 0} \int_a^b \frac{dx}{x^2} = \lim_{a\to 0}\left\{-\left(\frac{1}{b}-\frac{1}{a}\right)\right\}$$

which is infinite.

The notation $\int_0^b \frac{dx}{x^2}$ is used to mean $\lim_{a\to 0}\int_a^b \frac{dx}{x^2}$

Thus $\int_0^b \frac{dx}{x^2}$ is infinite, and the integral does not exist.
To summarize:

▶ $\int_a^\infty \frac{dx}{x^2}$ exists and has the value $\frac{1}{a}$

▶ $\int_0^b \frac{dx}{x^2}$ does not exist.

It is clear therefore that in all such cases you must investigate and determine whether the definite integral has a finite value or not.

Here is an example in which both limits become infinite.

$$\int_a^b \frac{dx}{1+x^2} = [\tan^{-1} x]_a^b$$

$$= \tan^{-1} b - \tan^{-1} a$$

If $b \to \infty$, then $\tan^{-1} b \to \frac{1}{2}\pi$

If $a \to \infty$, then $\tan^{-1} a \to -\frac{1}{2}\pi$

Therefore, in the limit

$$\int_{-\infty}^{\infty} \frac{dx}{1+x^2} \text{ becomes } \left\{ \frac{1}{2}\pi - \left(-\frac{1}{2}\pi \right) \right\} = \pi$$

Therefore the integral $\int_{-\infty}^{\infty} \frac{dx}{1+x^2}$ has a finite value.

14.7 Functions with infinite values

There are some functions which become infinite for some value, or values, of the variable between the limits of the definite integral, that is, the function is not continuous.

The function $\frac{1}{x^2}$ (considered in Section 14.6 above) is an example. It becomes infinite when $x = 0$ (as shown above). If therefore you need to find the value of the integral $\int_{-2}^{2} \frac{dx}{x^2}$ you can see that the function becomes infinite for a value of x between the limits, namely $x = 0$.

If you try to evaluate $\int_{-2}^{2} \frac{dx}{x^2}$ as usual, disregarding this infinite value, the result is as follows.

$$\int_{-2}^{2} \frac{dx}{x^2} = \left[-\frac{1}{x} \right]_{-2}^{2} = -\left\{ \frac{1}{2} - \left(-\frac{1}{2} \right) \right\} = -1$$

But this result contradicts the result of Section 14.6 above when it was shown that $\int_{-2}^{2} \frac{dx}{x^2}$ is infinite.

As $\int_{-2}^{2} \frac{dx}{x^2} = \int_{-2}^{0} \frac{dx}{x^2} + \int_{0}^{2} \frac{dx}{x^2}$ (see Section 14.4) and from the symmetry of the graph in Figure 14.5 $\int_{-2}^{0} \frac{dx}{x^2} = \int_{0}^{2} \frac{dx}{x^2}$ the integral $\int_{-2}^{2} \frac{dx}{x^2}$ must be infinite.

It is therefore necessary, before evaluating integrals, to ascertain if the function is continuous between the assigned limits, or whether it becomes infinite for some value of x.

This is especially necessary in the case of fractional functions in which, while the numerator remains finite, the denominator vanishes for one or more values of x.

Thus $\frac{x}{(x-1)(x-2)}$ becomes infinite, and the curve is discontinuous:

- when $x - 1 = 0$, that is $x = 1$
- when $x - 2 = 0$, that is $x = 2$

Similarly in $\frac{1}{\sqrt{2-x}}$ the denominator vanishes when $x = 2$, or more accurately, $\sqrt{2-x} \to 0$ when $x \to 2$. Consequently the function approaches infinity as $x \to 2$.

You must examine all such functions individually to ascertain if a finite limit and therefore a definite value of the integral exists. For this purpose you can often use the property of an integral as stated in Section 14.4. In using this theorem, the integral to be tested is expressed as the sum of two integrals in which the value of the variable for which the function becomes infinite is used as an end limit. Each of these must have a finite value if the original integral is finite and its value is given by that sum.

An example of this was given above. It was shown that $\int_{-2}^{2} \frac{dx}{x^2}$, when expressed as the sum of $\int_{-2}^{0} \frac{dx}{x^2}$, and $\int_{0}^{2} \frac{dx}{x^2}$, must be infinite, that is, it has no meaning, since each of the two component integrals had been shown to be infinite. A further example is given in Example 14.7.1 below in which a method is employed for determining whether a given definite integral is infinite or not.

Example 14.7.1

Determine whether the definite integral $\int_{0}^{3} \frac{dx}{(x-2)^{\frac{1}{3}}}$ has a finite value.

The function to be integrated approaches infinity as $x \to 2$.

Using the result in Section 14.4, you can express the integral in the form

$$\int_0^3 \frac{dx}{(x-2)^{\frac{1}{3}}} = \int_0^2 \frac{dx}{(x-2)^{\frac{1}{3}}} + \int_2^3 \frac{dx}{(x-2)^{\frac{1}{3}}}$$

It is necessary, if the original integral is to have a finite value that each of these integrals should be finite. Therefore, each of them must be tested separately.

In the first let the end limit '2' be replaced by $2 - \alpha$ (where α is a small positive number).

1 Then

$$\int_0^{2-\alpha} \frac{dx}{(x-2)^{\frac{1}{3}}} = \frac{3}{2}\left[(x-2)^{\frac{2}{3}}\right]_0^{2-\alpha}$$

$$= \frac{3}{2}\left[\{(2-\alpha)-2\}^{\frac{2}{3}} - (0-2)^{\frac{2}{3}}\right]$$

$$= \frac{3}{2}\left\{(-\alpha)^{\frac{2}{3}} - (-2)^{\frac{2}{3}}\right\}$$

As $\alpha \to 0$, $(-\alpha)^{\frac{2}{3}} \to 0$ and the value of the integral approaches

$$\frac{3}{2}(-2)^{\frac{2}{3}} = -\frac{3}{2}(2)^{\frac{2}{3}}$$

Similarly

$$\int_{2+\alpha}^3 \frac{dx}{(x-2)^{\frac{1}{3}}} = \frac{3}{2}\left[(x-2)^{\frac{2}{3}}\right]_{2+\alpha}^3$$

$$= \frac{3}{2}\left[\{(3-2)^{\frac{2}{3}}\} - \{(2+\alpha)-2\}^{\frac{2}{3}}\right]$$

$$= \frac{3}{2}\left(1-\alpha^{\frac{2}{3}}\right)$$

$$= \frac{3}{2} - \frac{3}{2}\alpha^{\frac{2}{3}}$$

As $\alpha \to 0$, and $2 + \alpha \to 2$, the value of the integral approaches $\frac{3}{2}$. As each of the definite integrals has a finite value, the whole integral is finite and is equal to the sum of the two integrals.

Therefore $\int_0^3 \frac{dx}{(x-2)^{\frac{1}{3}}} = \frac{3}{2} - \frac{3}{2}(2)^{\frac{2}{3}} = \frac{3}{2}\left(1 - 2^{\frac{2}{3}}\right)$

Exercise 14.2

Where possible, calculate the values of the following definite integrals.

1 $\int_a^\infty \frac{dx}{x}$

2 $\int_2^\infty \frac{dx}{x^3}$

3 $\int_0^\infty \frac{dx}{x^2+1}$

4 $\int_2^\infty \frac{dx}{x^2-1}$

5 $\int_2^\infty \frac{dx}{\sqrt{x}}$

6 $\int_0^\infty e^{-x}\,dx$

7 $\int_0^\infty e^{-x}\cos x\,dx$

8 $\int_0^\infty \frac{x\,dx}{1+x^2}$

9 $\int_1^\infty \frac{dx}{x^2(1+x)}$

10 $\int_1^\infty \frac{dx}{x(1+x)^2}$

11 $\int_0^1 \frac{dx}{x}$

12 $\int_0^1 \sqrt{\frac{x}{1-x}}\,dx$

13 $\int_{-1}^1 \sqrt{\frac{1-x}{1+x}}\,dx$

14 $\int_0^1 x\ln x\,dx$

15 $\int_{-1}^1 \frac{1+x}{1-x}\,dx$

16 $\int_0^\infty x^2 e^{-x}\,dx$

17 $\int_0^1 \ln x\,dx$

18 $\int_0^1 \frac{dx}{\sqrt{1-x}}$

19 $\int_0^3 \frac{dx}{(x-1)^2}$

20 $\int_0^2 \frac{dx}{(x-1)^{\frac{1}{3}}}$

Key ideas

▶ Just as the derivative of a function tells you about the gradient of the curve, its integral tells you about the area under the curve.

▶ The term 'area under the curve' should be specified more clearly; it is the area bounded by the curve, the x-axis and two vertical lines (or ordinates) usually denoted by $x = a$ and $x = b$.

▶ An integral where the limits are specified is a *definite integral* and it can normally be calculated to give a 'definite' numerical answer.

▶ If the limits of a definite integral are swapped over, the numerical value of the integral changes sign.

▶ Care must be taken when integrating functions that contain infinite values as integration will only give sensible answers where the definite integral is finite.

15

The integral as a sum; areas

In this chapter you will learn:

- ▶ *how to think of an integral as a limit of a sum*
- ▶ *to find new formulae which involve integration for finding area*
- ▶ *how to use polar coordinates and find area in the context of polar coordinates.*

15.1 Approximation to area by division into small elements

In the preceding chapter you saw how, using integration, you can find the area of a figure bounded in part by a curve of which the equation is known. Here is another, more general, treatment of the problem.

In Figure 15.1, let AB be a portion of a curve with equation $y = \phi(x)$. Let AM, BN be the ordinates of A and B, so that $OM = a$, $ON = b$, and $AM = \phi(a)$. $BN = \phi(b)$. $ABNM$ is the figure whose area is required. Let MN be divided into n equal parts at $X_1 X_2$, X_3 ... . Then let $MX_1 = \delta x$. Hence each of the divisions $X_1 X_2$, $X_2 X_3$, ... is equal to δx.

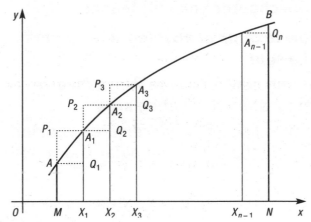

Figure 15.1

Let $A_1 X_1$, $A_2 X_2$, $A_3 X_3$ ... be ordinates corresponding to the points A_1, A_2, A_3, ...

Complete the rectangles $AP_1 A_1 Q_1$, $A_1 P_2 A_2 Q_2$, $A_2 P_3 A_3 Q_3$, ...

There are now two sets of rectangles corresponding to the divisions MX_1, $X_1 X_2$, ...

1 $MP_1 A_1 X_1$, $X_1 P_2 A_2 X_2$, $X_2 P_3 A_3 X_3$, ...

2 $MAQ_1 X_1$, $X_1 A_1 Q_2 X_2$, $X_2 A_2 Q_3 X_3$, ...

The area beneath the curve, that is, the area of $MABN$, lies between the sums of the areas of the rectangles in the sets **1** and **2**. If you increase the number of divisions the area of each of the two sets approximates more closely to the area of $MABN$. If you increase the number of divisions indefinitely, δx will decrease indefinitely and the area of each of the sets **1** and **2** approaches the area under the curve. You therefore need to find expressions for the sums of these sets and then to obtain their limiting values when $\delta x \to 0$.

The y-values for the points on the curve are:

$$AM = \phi(a)$$
$$A_1 X_1 = \phi(a + \delta x)$$
$$A_2 X_2 = \phi(a + 2\delta x)$$
$$\vdots$$
$$A_{n-1} X_{n-1} = \phi(a + (n-1)\delta x)$$
$$BN = \phi(b)$$

where n is the number of divisions. Therefore the areas of the rectangles in set **2** are as follows:

$$\text{Area } MAQ_1 X_1 = (AM \times MX_1) = \phi(a)\delta x$$
$$\text{Area } X_1 A_1 Q_2 X_2 = (A_1 X_1 \times X_1 X_2) = \phi(a + \delta x)\delta x$$
$$\text{Area } X_2 A_2 Q_3 X_3 = (A_2 X_2 \times X_2 X_3) = \phi(a + 2\delta x)\delta x$$
$$\vdots$$
$$\text{Area } X_{n-1} A_{n-1} Q_n N = (A_{n-1} X_{n-1} \times X_{n-1} N) = \phi(a + (n-1)\delta x)\delta x$$

The sum of all these rectangles is:

$$\delta x[\phi(a) + \phi(a + \delta x) + \cdots + \phi(a + (n-1)\delta x)] \qquad \textbf{A}$$

Similarly the sum of all the rectangles in set **1** is:

$$\delta x[\phi(a + \delta x) + \phi(a + 2\delta x) + \cdots + \phi(a + (n-1)\delta x) + \phi(b)] \qquad \textbf{B}$$

The area of the figure $AMNB$ lies between **A** and **B**

Then
$$\textbf{B} - \textbf{A} = \delta x(\phi(b) - \phi(a))$$

In the limit as $\delta x \to 0$ this difference vanishes. Thus each of the areas approaches the area of $AMNB$. Therefore the area is the limit of the sum of either **A** or **B**.

The summation of such a series can be expressed concisely by the use of the symbol Σ (pronounced 'sigma'), the Greek capital 'S'. Using this symbol the sum of the series may be written:

$$\sum_{x=a}^{x=b} \phi(x)\delta x$$

This expression means the sum of terms of the type $\phi(x)\delta x$, when you substitute for x the values:

$$a, a + \delta x, a + 2\delta x, a + 3\delta x, \ldots$$

for all such possible values between $x = a$ and $x = b$.

The area of $AMNB$ is the limit of this sum as $\delta x \to 0$. This is written in the form:

$$A = \lim_{\delta x \to 0} \sum_{x=a}^{x=b} \phi(x)\delta x$$

But you have seen, in Section 14.2, that this area is given by the integral:

$$\int_a^b \phi(x)\,dx$$

Therefore

$$\int_a^b \phi(x)\,dx = \lim_{\delta x \to 0} \sum_{x=a}^{x=b} \phi(x)\delta x$$

15.2 The definite integral as the limit of a sum

A definite integral can therefore be regarded as a sum, or, more correctly, as a 'limit of a sum', of the areas of a large number of very thin rectangles of width δx.

Nugget – The limit of a sum

Pause for a moment and consider what is meant here by the phrase 'the limit of a sum'. As you have seen above, we started thinking about finding the area under a curve by considering the discrete case where the area was divided into a finite number of narrow rectangles, each of width δx. The notation used to indicate the summation of these areas was the Greek letter 'Sigma', Σ. You then were invited to imagine gradually narrowing the width of each rectangle and thereby increasing their number. In the limit, the width of each rectangle reduces to zero and their number becomes infinite. This shift from the discrete to the continuous case is signalled by a change in notation: the δx is replaced by dx and the Σ is replaced by $\int$.

Note that it is this shift from the discrete to the continuous case that is implied in the phrase, 'the limit of a sum'.

The use of the term 'integral' will now be clear. The word 'integrate' means 'to give the total sum'. The first letter of the word sum appears in the sign $\int$ which is the old-fashioned elongated 's'. It is also evident why the dx appears as a factor in an integral.

The definite integral has been used in the illustration above to refer to the sum of areas. This familiar geometrical example, however, is a device for illustrating the process of summation. Actually what was found was the sum of an infinite number of algebraic products, one factor of which, in the limit, becomes infinitely small. The results, however, can be reached independently of any geometrical illustration.

Consequently, you can think of $\int_a^b \phi(x)\,dx$ as the limit of a sum of a very large number of products, one factor of which is a very small quantity. The successive products must be of the nature of those appearing in the demonstration above, and must refer to successive values of the independent variable, x, between the limits $x = a$ and $x = b$.

When this happens, the method can be applied to summing any such series, subject to the conditions which have been stated. This is of great practical importance. It enables you to calculate not only areas, but also volumes, lengths of curves, centres of mass, moments of inertia, etc., expressed in the form

$$\sum_{x=a}^{x=b} \phi(x)\delta x$$

They can then be represented by the definite integral $\int_a^b \phi(x)\,dx$

In the above demonstration $\phi(x)$ has been regarded as steadily and continuously increasing, but it is possible to modify the argument to apply to cases when that is not so. There are many practical applications of this process; some of them are discussed in the following chapters. The most obvious application, in view of the method followed in this demonstration, is to areas: so it would seem sensible for the next stage of the study to be made by examining some examples of them.

15.3 Examples of areas

Example 15.3.1

Find the area between the graph of $y = \frac{1}{2}x^2$, the x-axis, and the straight line $x = 2$.

The part of the curve involved is shown in Figure 15.2 (below) by OQ, where the ordinate from Q corresponds to the point $x = 2$.

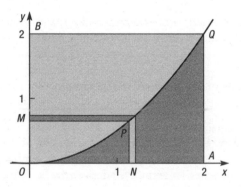

Figure 15.2

The area required is that of the region OAQ indicated by dark shading.

Let $P(x, y)$ be any point on the curve, so that $ON = x$. Let x be increased by δx. Then, drawing the corresponding ordinate, there is enclosed what is approximately a small rectangle, as shown in the figure.

The area of this 'rectangle' is approximately $y\delta x$.

When δx becomes very small, the sum of the areas of all such rectangles throughout the range from $x = 0$ to $x = 2$ is equal to the required area.

This thin rectangle is called an **element of area**; it is always necessary to obtain such an element before proceeding to the solution.

The sum of all such areas is given by the definite integral:

$$\int_0^2 y\,dx$$

But $\qquad\qquad y = \tfrac{1}{2}x^2$

Therefore the area $= \int_0^2 \tfrac{1}{2}x^2\,dx = \tfrac{1}{2}\left[\tfrac{1}{3}x^3\right]_0^2 = 1\tfrac{1}{3}$ square units.

Example 15.3.2

Find the area between the graph of $y = \tfrac{1}{2}x^2$, the y-axis and the straight line $y = 2$.

The curve is the same as in Example 15.3.1, and is shown in Figure 15.2 with light shading; BQ is the line $y = 2$.

Take any point $P(x, y)$ on the curve; as before, $OM = y$ and $ON = x$.

In this problem it is convenient to consider the area of the dark, thin 'rectangle' of thickness δy and length x. Then the area of this 'rectangle' is approximately $x\delta y$.

When δy becomes indefinitely small, the sum of the areas of all such rectangles throughout the range from $y = 0$ to $y = 2$ is equal to the required area. The area $OBQO$ is given by the definite integral:

$$\int_{y=0}^{y=2} x\,dy$$

There are two variables in the integral, and you must express one of them in terms of the other so that there remains one variable only. Express x in terms of y, so that the limits are unaltered.

Since $y = \frac{1}{2}x^2$ and $x \geq 0$, $x = \sqrt{2y}$. Substituting:

$$\text{area} = \int_0^2 \sqrt{2y}\,dy = \sqrt{2}\int_0^2 y^{\frac{1}{2}}\,dy = \sqrt{2}\left[\frac{2}{3}y^{\frac{3}{2}}\right]_0^2$$
$$= \sqrt{2} \times \frac{2}{3} \times 2^{\frac{3}{2}} = 2\frac{2}{3} \text{ square units}$$

If dy had been expressed in terms of x, so that $dy = x\,dx$, then you would have to change the limits. You would need to obtain the values of x corresponding to $y = 0$ and $y = 2$. In this case they are the same, since from $y = \frac{1}{2}x^2$, when $y = 0$, $x = 0$, and when $y = 2$, $x = 2$.

Notice that the sum of this area and the area in Example 15.3.1 must equal the area of the rectangle $OBQA$, which is

$$2\frac{2}{3} + 1\frac{1}{3} = 4 \text{ square units}$$

Example 15.3.3

Find the area of a circle.

1 Area by rectangular coordinates

Before finding the area enclosed wholly or in part by a curve, you need to know the equation of that curve.

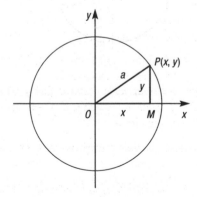

Figure 15.3

In any circle, the centre can be taken as the origin of a system of coordinates, and two diameters at right angles to each other as the coordinate axes. This is indicated in Figure 15.3.

Take any point $P(x, y)$ on the circumference of the circle and draw PM perpendicular to the x-axis.

Let the radius of the circle be a

Then $OM = x$ and $MP = y$

By the property of a right-angled triangle

$$OM^2 + MP^2 = OP^2$$

that is
$$x^2 + y^2 = a^2$$

This equation is true for any point on the circumference, and it states the relation which exists between the coordinates of any point and the constant which defines the circle, that is, the radius a. Therefore $x^2 + y^2 = a^2$ is the equation of a circle of radius a and the origin at its centre. Figure 15.4 shows this circle.

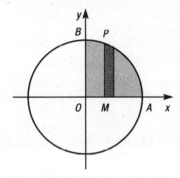

Figure 15.4

For reasons which will be apparent later in Section 15.4, it is better to find the area of the quadrant which is shaded; from this you can find the area of the whole circle.

Let $P(x, y)$ be any point on the circumference. Then $OM = x$ and $MP = y$

The element of area, as previously defined, has an approximate area of $y\delta x$.

For the purpose of the definite integral which gives you the area, the limits of x for the quadrant are at O, $x = 0$, and at A, $x = a$.

Therefore
$$\text{area} = \int_0^a y\, dx$$

Since $x^2 + y^2 = a^2$ and $y \geq 0$

$$y = \sqrt{a^2 - x^2}$$

and
$$\text{area} = \int_0^a \sqrt{a^2 - x^2}\, dx$$

In Section 12.4 it was shown that:

$$\int \sqrt{a^2 - x^2}\, dx = \tfrac{1}{2}a^2 \sin^{-1} \tfrac{x}{a} + \tfrac{1}{2}x\sqrt{a^2 - x^2} + C$$

Therefore
$$\text{area} = \int_0^a \sqrt{a^2 - x^2}\, dx$$
$$= \left[\tfrac{1}{2}a^2 \sin^{-1} \tfrac{x}{a} + \tfrac{1}{2}x\sqrt{a^2 - x^2} \right]_0^a$$

Now when $\quad x = a, \sin^{-1}\tfrac{x}{a} = \sin^{-1} 1 = \tfrac{1}{2}\pi$

and when $\quad x = 0, \sin^{-1}\tfrac{x}{a} = \sin^{-1} 0 = 0$

Therefore
$$\text{area} = \left\{ \tfrac{1}{2}a^2 \times \tfrac{1}{2}\pi + \tfrac{1}{2}a\sqrt{a^2 - a^2} \right\} - 0$$
$$= \tfrac{1}{4}\pi a^2$$

Therefore, the area of the circle $= \pi a^2$

2 Alternative method

You may find the following method useful in applications. You can think of the area of a circle as the area of a plane figure which is traced out by a finite straight line as it rotates around one of its ends, and makes a complete rotation.

Thus in Figure 15.5 if the straight line OP, length a units, starting from the fixed position OA on the x-axis, makes a complete rotation around the fixed point O, the point P describes the circumference of a circle, and the area marked out by OA is the area of the circle.

Let the point P have rotated from the x-axis, so that it has described the angle, θ. Then AOP is a sector of a circle.
Now suppose OP rotates further through a small angle $\delta\theta$.

The resulting small sector is an element of area, and the sum of all such sectors when OP makes a complete rotation from the x-axis, back again to its original position, will be the area of the circle.

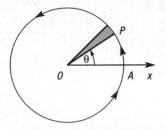

Figure 15.5

You can think of the small arc subtended by $\delta\theta$ as a straight line, and the small sector as a triangle.

The length of the arc is $a\,\delta\theta$.

If $\delta\theta$ is small, the altitude of the triangle is very close to the radius, a, of the circle.

Using the formula for the area of a triangle,

element of area $= \frac{1}{2} \times$ base $\times$ height $= \frac{1}{2} \times a\,\delta\theta \times a = \frac{1}{2}a^2\,\delta\theta$

The angle corresponding to a complete rotation is 2π radians.

Therefore
$$\text{area} = \int_0^{2\pi} \tfrac{1}{2}a^2\,d\theta = \left[\tfrac{1}{2}a^2\theta\right]_0^{2\pi}$$
$$= \tfrac{1}{2}a^2 \times 2\pi = \pi a^2 \text{ square units}$$

Example 15.3.4

Find the area of part of a circle between two parallel chords.

In the circle $x^2 + y^2 = 9$ find the area contained between the lines $x = 1$ and $x = 2$. The radius of this circle is 3 and its centre is at the origin. The area which you are required to find is shown in Figure 15.6 (below).

Since $x^2 + y^2 = 9$, $y = \sqrt{9 - x^2}$ for the part of the circle above the x-axis. The area of the region above the x-axis is given by

$$\text{area} = \int_1^2 y\,dx = \int_1^2 \sqrt{9 - x^2}\,dx$$

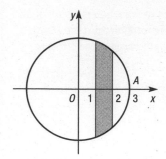

Figure 15.6   **Figure 15.7**

Using the result from Section 12.4:

$$\int \sqrt{a^2 - x^2}\, dx = \tfrac{1}{2}a^2 \sin^{-1}\frac{x}{a} + \tfrac{1}{2}x\sqrt{a^2 - x^2} + C$$

$$\int_1^2 \sqrt{9 - x^2}\, dx = \left[\tfrac{9}{2}\sin^{-1}\tfrac{1}{3}x + \tfrac{1}{2}x\sqrt{9 - x^2} \right]_1^2$$

$$= \left\{ \tfrac{9}{2}\sin^{-1}\tfrac{2}{3} + \tfrac{1}{2} \times 2\sqrt{9 - 4} \right\} - \left\{ \tfrac{9}{2}\sin^{-1}\tfrac{1}{3} + \tfrac{1}{2}\sqrt{9 - 1} \right\}$$

$$= \left\{ \tfrac{9}{2}\sin^{-1}\tfrac{2}{3} + \sqrt{5} \right\} - \left\{ \tfrac{9}{2}\sin^{-1}\tfrac{1}{3} + \tfrac{1}{2}\sqrt{8} \right\}$$

$$= \sqrt{5} - \sqrt{2} + \tfrac{9}{2}\left\{ \sin^{-1}\tfrac{2}{3} - \sin^{-1}\tfrac{1}{3} \right\}$$

This result is the area above the x-axis. The total area both above and below the x-axis, is double this, that is:

$$\text{area} = 2\sqrt{5} - 2\sqrt{2} + 9\left\{ \sin^{-1}\tfrac{2}{3} - \sin^{-1}\tfrac{1}{3} \right\}$$

This area is approximately 5.153 square units, to 3 decimal places.

Example 15.3.5

Find the area of the segment cut off from the circle $x^2 + y^2 = 9$ by the line $x = 2$. This is the same circle as in the previous example, and the area required is shaded in Figure 15.7. Considering only the area of that part lying above the x-axis you find:

$$\text{area} = \int_2^3 y\, dx = \int_2^3 \sqrt{9 - x^2}\, dx$$

so $\quad$ total area $= 2\int_2^3 y\,dx = 2\int_2^3 \sqrt{9-x^2}\,dx$

Using the result obtained in the Example 15.3.4:

$$2\int_2^3 \sqrt{9-x^2}\,dx = 2\left[\tfrac{9}{2}\sin^{-1}\tfrac{1}{3}x + \tfrac{1}{2}x\sqrt{9-x^2}\right]_2^3$$

$$= 2\left\{\tfrac{9}{2}\sin^{-1}\tfrac{3}{3} + \tfrac{1}{2}\times 3\sqrt{0}\right\} - 2\left\{\tfrac{9}{2}\sin^{-1}\tfrac{2}{3} + \tfrac{1}{2}\times 2\sqrt{5}\right\}$$

$$= 2\left\{\tfrac{9}{2}\sin^{-1}1 + 0\right\} - 2\left\{\tfrac{9}{2}\sin^{-1}\tfrac{2}{3} + \sqrt{5}\right\}$$

$$= \tfrac{9}{2}\pi - 9\sin^{-1}\tfrac{2}{3} - 2\sqrt{5}$$

This area is approximately 3.097 square units, to 3 decimal places. Notice, as a check, that you should find the area of the segment cut off by the line $x = 1$. It should be the sum of those above.

Example 15.3.6

Find the area of the ellipse $\dfrac{x^2}{a^2} + \dfrac{y^2}{b^2} = 1$ shown in Figure 15.8.

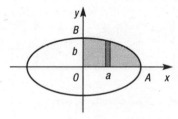

Figure 15.8 The ellipse $\dfrac{x^2}{a^2} + \dfrac{y^2}{b^2} = 1$

The area of the element is $y\delta x$, so the area of the shaded region is

$$\int_0^a y\,dx$$

From the equation $\dfrac{x^2}{a^2} + \dfrac{y^2}{b^2} = 1,\ y = \dfrac{b}{a}\sqrt{a^2 - x^2}$

Therefore the total area of the ellipse is:

$$4\int_0^a y\,dx = 4\int_0^a \tfrac{b}{a}\sqrt{a^2 - x^2}\,dx$$

$$= 4\tfrac{b}{a}\int_0^a \sqrt{a^2 - x^2}\,dx$$

$$= 4\tfrac{b}{a}\left[\tfrac{1}{2}a^2\sin^{-1}\tfrac{x}{a} + \tfrac{1}{2}x\sqrt{a^2 - x^2}\right]_0^a$$

$$= 4\tfrac{b}{a}\times\tfrac{1}{2}a^2\times\tfrac{1}{2}\pi = \pi ab$$

If you compare this with the area of the circle, radius a, in Example 15.3.3, you see that the ratio of the area of the quadrant of the ellipse to that of the corresponding area of the circle of radius a is $\frac{b}{a}$, that is, the ratio of the minor axis to the major. This is also the ratio of corresponding ordinates of the two curves.

Example 15.3.7

Find the area enclosed between the curve of $y = \frac{4}{x+1}$, the x-axis and the lines $x = 1$ and $x = 4$. The graph of this function is a hyperbola and shown in Figure 15.9.

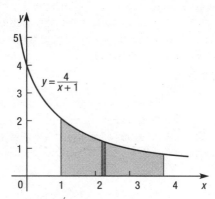

Figure 15.9 Graph of $y = \frac{4}{x+1}$

The area you are required to find is shaded.

Taking the element of area as $y\delta x$ and substituting $y = \frac{4}{x+1}$

$$\text{area} = \int_1^4 \frac{4}{x+1}dx$$
$$= 4[\ln(x+1)]_1^4$$
$$= 4\ln 5 - 4\ln 2$$
$$= 4\ln\frac{5}{2}$$

15.4 Sign of an area

Nugget – Positive and negative areas

Just as numbers can take both positive and negative values, so too can areas. Areas above the x-axis are defined as positive, while areas below are negative. Unfortunately, integration gives only positive answers for areas, regardless of whether they lie above or below the x-axis. As you will see in this section, this complication requires us to identify these regions and calculate each of their areas separately, after which we can make the adjustment for whether they are positive or negative.

So far in the examples of calculating areas, the regions concerned were, in most cases, lying above the x-axis, and the values of the function were consequently positive. In the examples of the circle and ellipse, in which the curve is symmetrical about both axes, positive values of the function were still used by finding the area in one quadrant and then multiplying by 4. Areas which lie below the x-axis, where the values of the function are negative, are now considered.

Example 15.4.1

Find the area enclosed between the curve $y = x^2 - 3x + 2$ and the x-axis.

Since $x^2 - 3x + 2 = (x - 1)(x - 2)$ the curve cuts the x-axis at $x = 1$ and $x = 2$.

As $\frac{dy}{dx} = 2x - 3$, there is a turning point when $\frac{dy}{dx} = 2x - 3 = 0, x = 1.5$

Since $\frac{d^2y}{dx^2} = 2$ and is always positive, this point is a minimum.

The curve is shown in Figure 15.10 (below); the area required lies entirely below the x-axis.

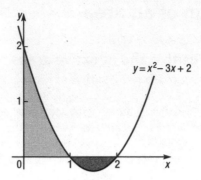

Figure 15.10

Let this area be A.

Then
$$A = \int_1^2 (x^2 - 3x + 2)\,dx$$
$$= \left[\tfrac{1}{3}x^3 - \tfrac{3}{2}x^2 + 2x\right]_1^2$$
$$= \left(\tfrac{8}{3} - 6 + 4\right) - \left(\tfrac{1}{3} - \tfrac{3}{2} + 2\right)$$
$$= -\tfrac{1}{6}$$

The result is a negative area. But an area, fundamentally, is signless. How, then, should you interpret this result? It will probably not come as a surprise because you will have seen that the definite integral $\int_a^b y\,dx$ represents the sum of a large number of elements of area of the form $y\delta x$ which are themselves small. When the area lies below the x-axis, all the values of the function, that is, y, are negative, and since δx which represents an increase, is positive, all the products $y\,\delta x$ must be negative. Hence the sum is negative. It was shown in Section 15.2 that the summation is general for all such products, and the representation of an area by it is just one of the applications. Hence if you are finding an actual area by the integration, the negative sign must be disregarded. Since by the convention of signs used in the graphical representation of a function ordinates below the axis are negative, the corresponding areas are also negative. Hence as a matter of convention, areas above the x-axis are considered positive and below the axis are negative.

Note also the following points in connection with Example 15.4.1.

▶ The area below the curve between $x = 0$ and $x = 1$, that is, the lighter shaded area Figure 15.10, is given by

$$\int_0^1 (x^2 - 3x + 2)dx = \tfrac{5}{6}$$

▶ Consequently the total shaded area disregarding the negative sign of the integral $\int_0^2 (x^2 - 3x + 2)dx$, is $\tfrac{5}{6} + \tfrac{1}{6} = 1$

▶ The total area as given by the integral

$$\int_0^2 (x^2 - 3x + 2)dx = \tfrac{2}{3}$$

that is $\tfrac{5}{6} - \tfrac{1}{6}$

Example 15.4.2

Find the area enclosed between the curve $y = 4x(x - 1)(x - 2)$ and the x-axis.

Figure 15.11 shows the relevant part of the graph. The areas required are shaded.

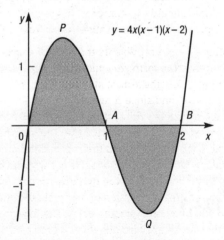

Figure 15.11

As the function $4x(x - 1)(x - 2)$ vanishes for $x = 0$, 1 and 2 its graph cuts the x-axis for these values of x. Using the method of Section 7.4 shows that there are two turning points, a maximum value of 1.55 when $x = 0.45$; and a minimum value of -1.55 when $x = 1.55$. You calculate the area in two parts.

$$\text{Area of } OPA = \int_0^1 4x(x-1)(x-2)\,dx$$

$$= \int_0^1 (4x^3 - 12x^2 + 8x)\,dx$$

$$= \left[x^4 - 4x^3 + 4x^2 \right]_0^1$$

$$= (1 - 4 + 4) - (0)$$

$$= 1 \text{ square unit}$$

$$\text{Area of } AQB = \int_1^2 4x(x-1)(x-2)\,dx$$

$$= \left[x^4 - 4x^3 + 4x^2 \right]_1^2$$

$$= -1 \text{ square unit}$$

Hence, disregarding the negative sign of the lower area, the total area of the shaded regions is 2 square units.

If you integrate for the whole area between the limits 0 and 2 you get:

$$\text{area} = \int_0^2 4x(x-1)(x-2)\,dx$$

$$= \left[x^4 - 4x^3 + 4x^2 \right]_0^2$$

$$= 16 - 32 + 16$$

$$= 0$$

This agrees with the algebraic sum of the two areas found separately.

Examples 15.4.1 and 15.4.2 show that when you want to find the total area enclosed by the x-axis and a curve that crosses the axis, you must find separately the areas above and below the x-axis. Then the sum of these areas, disregarding the signs, will give the actual areas required.

Example 15.4.3

Find the area of the region enclosed between the x-axis and the graph of $y = \cos x$ between the following limits:

1 0 and $\frac{1}{2}\pi$ **2** $\frac{1}{2}\pi$ and π **3** 0 and π

1 The first region is the shaded area above the x-axis in Figure 15.12.

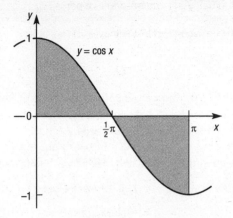

Figure 15.12

$$\text{Area} = \int_0^{\frac{1}{2}\pi} \cos x \, dx$$
$$= \left[\sin x \right]_0^{\frac{1}{2}\pi}$$
$$= \sin\tfrac{1}{2}\pi - \sin 0 = 1$$

2 The second area is shown with shading below the x-axis.

$$\text{Area} = \int_{\frac{1}{2}\pi}^{\pi} \cos x \, dx$$
$$= \left[\sin x \right]_{\frac{1}{2}\pi}^{\pi}$$
$$= \sin\pi - \sin\tfrac{1}{2}\pi$$
$$= 0 - 1 = -1$$

3 The third area is the sum of the first and second.

$$\int_0^{\pi} \cos x \, dx = \left[\sin x \right]_0^{\pi}$$
$$= \sin\pi - \sin 0$$
$$= 0$$

These results agree algebraically, but if you need to know the actual area between 0 and π, you must disregard the negative sign of the second area. The area of the two parts is therefore 2 square units.

Example 15.4.4

Find the area enclosed between the *x*-axis and the curve $y = \sin x$, for values of *x* between the following limits:

1 0 and π

2 π and 2π

From the part of the graph of $y = \sin x$ in Figure 15.13, you can see that the area enclosed between the curve and the *x*-axis consists of a series of loops of equal area, each corresponding to an interval of π radians, and lying alternately above and below the *x*-axis; consequently they are alternately positive and negative.

1 Area of the first loop $= \int_0^{\pi} \sin x \, dx = \left[-\cos x \right]_0^{\pi}$

$$= -(\cos \pi - \cos 0) = -\{(-1) - 1\} = 2$$

2 Area of the second loop $= \int_0^{2\pi} \sin x \, dx = \left[-\cos x \right]_0^{2\pi}$

$$= -(\cos 2\pi - (\cos \pi)) = -(1 - (-1)) = -2$$

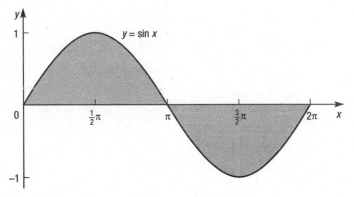

Figure 15.13

You can see that if there are *n* loops, where *n* is an odd integer, the total area, paying regard to the negative signs, is 2, but if *n* is even, the area calculated is zero.

The actual area of *n* loops is $2n$.

Example 15.4.5

Find the area contained between the graph of $y = x^3$ and the straight line $y = 2x$.

Figure 15.14 shows the graph of the functions between their points of intersection, A and A'. The area of the shaded regions is required.

You can see from symmetry that the parts above and below the x-axis will be equal in magnitude but opposite in sign.

Therefore you can find the area of $OABO$ and double it. This is the difference between the triangle OAC, and the area beneath the graph of $y = x^3$, namely, $OBAC$.

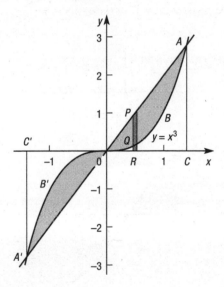

Figure 15.14

First find, as usual, an expression for the element of area.

From any point P on the line $y = 2x$ draw the ordinate PR, cutting the graph of $y = x^3$ in Q. As before, construct a small 'rectangle' which has PR as its left-hand edge, and width δx. This represents the element of area for the triangle, while the rectangle with QR as its left-hand edge represents the element of area for $OBAC$. Their difference, a rectangle with PQ as its left-hand edge represents the element of area for the shaded part.

Let $PR = y_1$ and $QR = y_2$

Then element of area with PR as its left-hand edge is $y_1\,\delta x$. Then element of area with QR as its left-hand edge is $y_2\,\delta x$. Therefore the element of area with PQ as its left-hand edge is $(y_1 - y_2)\delta x$. Before you can integrate, you must find the limits of the integral. These will be the value of x at O and A, the points of intersection of $y = x^3$ and $y = 2x$.

To find the value of x at these points solve the equations $y = x^3$ and $y = 2x$ simultaneously. Since $y = x^3$ and $y = 2x$, $2x = x^3$ so the roots are 0, $\sqrt{2}$ and $-\sqrt{2}$. These are values of x at O, A and A' respectively. For the positive area $OABO$ the limits are $x = 0$ and $x = \sqrt{2}$.

$$\text{Therefore the area required} = \int_0^{\sqrt{2}} (y_1 - y_2)dx$$
$$= \int_0^{\sqrt{2}} (2x - x^3)dx$$
$$= \left[x^2 - \tfrac{1}{4}x^4 \right]_0^{\sqrt{2}}$$
$$= (\sqrt{2})^2 - \tfrac{1}{4}(\sqrt{2})^4$$
$$= 1 \text{ square unit}$$

From symmetry and previous considerations you can conclude that the area below the x-axis is -1 square unit, which you can verify as follows:

$$A = \int_{-\sqrt{2}}^{0} (2x - x^3)dx$$
$$= \left[x^2 - \tfrac{1}{4}x^4 \right]_{-\sqrt{2}}^{0}$$
$$= 0 - \left[(\sqrt{2})^2 - \tfrac{1}{4}(\sqrt{2})^4 \right]$$
$$= -1 \text{ square unit}$$

Disregarding the negative sign the actual area of the two loops together is 2 square units. As an exercise you should verify this result by finding the area of the two loops using $\int_{-\sqrt{2}}^{\sqrt{2}} (2x - x^3)\,dx$

Exercise 15.1

Note: You are recommended to draw a rough sketch, using a graphics calculator if you wish, for each problem.

1 Find the area bounded by the graph of $y = x^3$, the x-axis and the lines $x = 2$ and $x = 5$.

2 Find the area bounded by the straight line $2y = 5x + 7$, the x-axis and the lines $x = 2$ and $x = 5$.

3 Find the area between the graph of $y = \ln x$, the x-axis and the lines $x = 1$, $x = 5$.

4 Find the area enclosed by the graph of $y = 4x^2$, the y-axis and the straight lines $y = 1$, $y = 4$.

5 Find the area between the graph of $y^2 = 4x$, the x-axis and the lines $x = 4$, $x = 9$.

6 Use integration to find the area of the circle $x^2 + y^2 = 4$.

7 In the circle $x^2 + y^2 = 16$ find the area included between parallel chords which are at distances of 2 and 3 units from the centre. Find the area of the segment cut off from the circle $x^2 + y^2 = 16$ by the chord which is 3 units from the centre.

8 Find by integration the area of the ellipse $\frac{x^2}{16} + \frac{y^2}{9} = 1$.

9 Find the area of the segment cut off from the hyperbola $\frac{x^2}{9} - \frac{y^2}{4} = 1$ by the chord $x = 4$.

10 Find the area between the hyperbola $xy = 4$, the x-axis and the ordinates $x = 2$, $x = 4$.

11 Find the area between the graph of $y = 2x - 3x^2$ and the x-axis.

12 Find the area bounded by $y = e^x$ and the x-axis between the ordinates $x = 0$ and $x = 3$.

13 Find the area cut off by the x-axis from the graph of $y = x^2 - x - 2$.

14 Find the whole area included between the graph of $y^2 = x^3$ and the line $x = 4$.

15 Find the area of the segment cut off from the curve of $xy = 2$ by the straight line $x + y = 3$.

16 Find the total area of the segments enclosed between the x-axis and the graph of $y = x(x-3)(x+2)$.

17 Find the area between the curves of $y = 8x^2$ and $y = x^3$.

18 Find the area between the two curves $y = x^2$ and $y^2 = x$.

19 Find the area between the catenary $y = \cosh \frac{1}{2}x$ (see Section 9.1), the x-axis and the ordinates $x = 0$, $x = 2$.

20 Find the area of the region between the graph of $y = x^2 - 8x + 12$, the x-axis and the ordinates $x = 1$, $x = 9$.

21 Find the actual area between the graph of $y = x^3$ and the straight line $y = \frac{1}{4}x$.

Nugget – Calculating areas above and below a curve using a machine

You may be wondering whether tackling these sorts of questions (i.e. with both positive and negative areas in the same calculation) can be done more easily on a calculator or suitable computer application. The good news here is that all well-designed applications will be able to do such calculations all in one go, treating both positive and negative areas correctly and summing these areas accordingly.

15.5 Polar coordinates

When you use polar coordinates instead of rectangular, you may find that equations of curves are sometimes simpler and finding areas may be easier. A brief account is given below. For a fuller treatment you will need to consult a text-book on coordinate geometry.

DEFINITIONS

Let OX (Figure 15.15) be a fixed straight line and O a fixed point on it. Then the position of any point P is defined with reference to these when you know:

▶ its distance from O

▶ the angle made by OP with OX.

Let r be this distance, and let θ be the angle made by OP with OX.

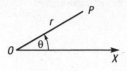

Figure 15.15   **Figure 15.16**

Then (r, θ) are called the **polar coordinates** of P. The fixed point O is called the **pole**, OP is called the **radius vector**, θ the **vectorial angle**, and OX the **initial line**. Notice that θ is the angle which would be described by the radius vector, in rotating in a positive direction from OX.

Connection between the rectangular coordinates and the polar coordinates of a point

Let P be a point (Figure 15.16) with polar coordinates (r, θ), and rectangular coordinates (x, y), namely OQ and PQ. Then you can see that $x = r \cos \theta$, $y = r \sin \theta$ and $x^2 + y^2 = r^2$.

POLAR EQUATION OF A CURVE

If a point moves along a curve, as θ changes, r in general will also change. Hence r is a function of θ.

The equation stating the relation between r and θ for a given curve is called the **polar equation** of the curve.

EXAMPLE OF A POLAR EQUATION

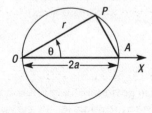

Figure 15.17

Let a point P move along the circumference of a circle (Figure 15.17). Let O be a fixed point at the extremity of a fixed diameter, and let the diameter of the circle in Figure 15.17 be $2a$. Then, for any position of P the polar coordinates are:

$$OP = r$$
$$\angle AOP = \theta$$

From elementary geometry $\angle OPA = 90°$. Therefore

$$r = 2a\cos\theta$$

This is the polar equation of the circle situated as described. If the centre of the circle were taken as the pole, r is always equal to a; that is, the polar equation is then $r = a$.

In this case r, the radius of the circle, is constant and independent of θ.

The equation of the circle may also take other forms.

15.6 Plotting curves from their equations in polar coordinates

The following example is typical of the method used to plot a curve from its polar equation.

Example 15.6.1

Draw the curve with polar equation $r = a(1 + \cos\theta)$.

The general method is to select values of θ, find the corresponding values of r, then plot the points obtained.

In Section 15.5, you saw that $r = a\cos\theta$ is the equation of a circle of diameter a, when the pole is on the circumference. Therefore for any value of θ, the value of r for the circle is increased by a; the result is the value of r for the required curve.

Draw a circle of radius $\frac{1}{2}a$ (Figure 15.18). Take a point O at the end of a diameter OA. O will be the pole for the curve.

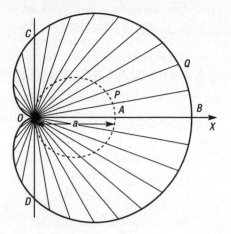

Figure 15.18

Since $\cos\theta$ takes its maximum value, 1, when $\theta = 0$, the maximum value of r for the curve will be at the point B, where $AB = a$.

Thus the maximum value of r is $2a$.

When $\theta = \frac{1}{2}\pi$ or $\frac{3}{2}\pi$, $\cos\theta = 0$ so $r = a$. These give the points C and D.

In the second quadrant $\cos\theta$ is decreasing to -1 when $\theta = \pi$, giving $r = a - a = 0$.

Similarly, the general shape of the curve may be found for the third and fourth quadrants.

Finally, when $\theta = 2\pi$, $\cos\theta = 1$ and the curve is closed at B.

To find other points on the curve between the points considered above, draw a series of chords of the circle, for increasing values of θ. If OP is one of these chords, produce it and mark off PQ equal to a. Then Q is a point on the curve. The complete curve, called the **cardioid**, from its heart-like shape, is shown in Figure 15.18. It is of importance in optics.

Other examples of equations of polar curves are

▶ the lemniscate, $r^2 = a^2 \cos 2\theta$

▶ the limacon, $r = b - a \cos\theta$

▶ the spiral of Archimedes, $r = a\theta$

▶ the logarithmic or equiangular spiral, $\ln r = a\theta$

▶ the hyperbolic spiral, $r\theta = a$

15.7 Areas in polar coordinates

In Figure 15.19, let AB be part of a curve for which the equation is known in polar coordinates.

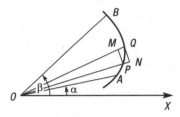

Figure 15.19

Suppose you want to find the area of the sector OAB contained between the curve and the two radii OA, OB, the angles made by them with the fixed line OX being:

$$\angle AOX = \alpha \text{ and } \angle BOX = \beta$$

Let P, with polar coordinates (r, θ), be any point on the curve so that $\angle POX = \theta$. Let Q be a neighbouring point on the curve, with polar coordinates $(r + \delta r, \theta + \delta\theta)$. Then, with the construction shown in Figure 15.19, the area of the sector OPQ lies between the areas of the triangles OPM and ONQ, the areas of which are:

$$\text{area } \triangle OPM = \tfrac{1}{2} r^2 \delta\theta \qquad \text{area } \triangle ONQ = \tfrac{1}{2}(r + \delta r)^2 \delta\theta$$

If the angle $\delta\theta$ is small, then δr will also be small, and the area of the sector is approximately $\tfrac{1}{2} r^2 \delta\theta$. This is, therefore, the element of area, and the sum of all such sectors between the limits $\theta = \alpha$ and $\theta = \beta$ will be the area of the sector OAB. Expressing this sum as an integral (as before):

$$\text{area of sector } OAB = \int_a^b \tfrac{1}{2} r^2 \, d\theta$$

When you know the polar equation of the curve, you can express r in terms of θ and evaluate the integral.

Example 15.7.1

Find the area of the circle with polar equation $r = 2a \cos\theta$ (see Section 15.5).

If P is a point moving round the curve, the radius vector OP describes the area of the circle.

When P is at A, $\theta = 0$; when P is at O, $\theta = \frac{1}{2}\pi$

As P moves from A to O, and the angle θ changes from 0 to $\frac{1}{2}\pi$, the area traced out by OP is a semicircle.

Using the formula, area of sector $OAB = \int_a^\beta \frac{1}{2}r^2\,d\theta$ (obtained above)

$$\text{area of semicircle} = \int_0^{\frac{1}{2}\pi} \frac{1}{2}r^2\,d\theta$$

So the area of the circle is twice this, and is given by

$$\text{area of circle} = \int_0^{\frac{1}{2}\pi} r^2\,d\theta$$

But $r = 2a\cos\theta$. Therefore

$$\begin{aligned}
\text{area of circle} &= \int_0^{\frac{1}{2}\pi} 4a^2\cos^2\theta\,d\theta \\
&= 4a^2 \int_0^{\frac{1}{2}\pi} \cos^2\theta\,d\theta \\
&= 4a^2 \int_0^{\frac{1}{2}\pi} \tfrac{1}{2}(1 + \cos 2\theta)d\theta \qquad \text{(Section 12.1)} \\
&= 2a^2 \left[\theta + \tfrac{1}{2}\sin 2\theta\right]_0^{\frac{1}{2}\pi} \\
&= 2a^2 \left(\tfrac{1}{2}\pi + \tfrac{1}{2}\sin\pi\right) \\
&= 2a^2 \times \tfrac{1}{2}\pi = \pi a^2
\end{aligned}$$

Exercise 15.2

1 Find the area of the cardioid with equation $r = a(1 + \cos\theta)$, the limits of θ being 0 and 2π.

2 Find the area of one loop of the curve $r = a\sin 2\theta$, that is, between the limits 0 and $\frac{1}{2}\pi$ How many loops are there between 0 and 2π?

Note: $a \sin 2\theta = 0$ when $\theta = 0$ and $\theta = \frac{1}{2}\pi$. As the function is continuous between these values, the curve must form a loop between them. You should draw roughly the whole curve.

3 Find the area of one loop of the lemniscate $r^2 = a^2 \cos 2\theta$. How many loops are there in the complete curve?

4 The radius vector of the function $r = a\theta$ makes one complete rotation from 0 to 2π. Find the area described.

5 Find the area which is described by the curve $r = a \sec^2 \frac{1}{2}\theta$ from $\theta = 0$ to $\theta = \frac{1}{2}\pi$.

6 Find the area enclosed by the curve $r = 3\cos\theta + 5$ between $\theta = 0$ and $\theta = 2\pi$.

15.8 Mean value

In Figure 15.20, let PQ be part of the graph of the function $y = f(x)$.

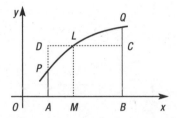

Figure 15.20

Let PA, QB be the ordinates at P and Q, where $OA = a$ and $OB = b$. Then the area of the region under the curve $APQB$ is given by:

$$\text{area of } APQB = \int_a^b f(x)\, dx.$$

Let $ABCD$ be a rectangle with area equal to that of $APQB$, that is to $\int_a^b f(x)\, dx.$ Let DC meet the curve at L.

Draw LM parallel to the y-axis, to meet the x-axis at M.

As the area of $ABCD = AB \times LM$, it *follows* that:

$$LM = \frac{\text{area of } ABCD}{AB}$$

$$= \frac{\text{area of } APQB}{AB}$$

$$= \frac{\int_a^b f(x)\,dx}{b-a}$$

LM is called the **mean value** of ordinates of the curve for the interval from $x = a$ to $x = b$. Therefore the mean value of $f(x)$ from $x = a$ to $x = b$ is:

$$\frac{\int_a^b f(x)dx}{b-a}$$

Example 15.8.1

Find the mean value of $2\cos t - \sin 3t$ between $t = 0$ and $t = \frac{1}{6}\pi$.

From the definition above, the mean value is given by:

$$\text{mean value} = \frac{\int_0^{\frac{1}{6}\pi} (2\cos t - \sin 3t)\,dt}{\frac{1}{6}\pi - 0}$$

$$= \frac{\left[2\sin t + \frac{1}{3}\cos 3t \right]_0^{\frac{1}{6}\pi}}{\frac{1}{6}\pi}$$

$$= \frac{1 - \frac{1}{3}}{\frac{1}{6}\pi}$$

$$= \frac{4}{\pi}$$

Exercise 15.3

1 Find the mean value of the function $\sin x$ over the range of values $x = 0$ to $x = \pi$.

2 Find the mean value of the function $\sin 2x$ over the range of values $x = 0$ to $x = \pi$.

3 Find the mean value of $y = \frac{1}{x}$ over the interval $x = 1$ to $x = 10$.

4 Find the mean value of $2\sqrt{x}$ between $x = 0$ and $x = 4$.

5 The equation of a curve is $y = b \sin^2 \frac{\pi x}{a}$. Find the mean height of the portion for which x lies between 0 and a.

6 Find the mean value of $\cos x$ between $x = 0$ to $x = \frac{1}{4}\pi$.

7 Find the mean value of the function $y = a \sin bx$ between the values $x = 0$ and $x = \frac{\pi}{b}$.

8 The range of a projectile fired with initial velocity v_0 and angle of elevation θ is $\frac{v_0^2}{g} \sin 2\theta$. Find the mean range as θ varies from 0 to $\frac{1}{2}\pi$.

Key ideas

▶ The word integral means 'total sum' and this is reflected in the sign $\int$, which is the old-fashioned, elongated letter 'S'.

▶ Conceptually, the problem of finding the area under a curve can be thought of as dividing the area into a large number of very thin rectangles and finding their total sum. In the limit, the number of these rectangles increases to infinity as their thickness reduces to zero.

▶ Just as the area between the curve and the x-axis can be calculated by finding the definite integral of *f(x) dx*, so the problem can be turned on its side and you can find the area between the curve and the y-axis by calculating the definite integral of *f(y) dy*.

▶ It is possible to find the area of a circle by the methods described earlier (i.e. calculating the definite integral of *f(x) dx*. An alternative approach, explained in Section 15.5, is to draw a radius and imagine it rotating through a small angle to create a tiny sector. The area of this sector can be found and its definite integral between the limits 0 and 2π will produce the area of the circle.

▶ Conventionally, areas above the x-axis are positive and those below the x-axis are negative. Unfortunately finding areas by integration gives only positive answers. Consequently, when finding the area enclosed between a curve and the x-axis, where some of the area is above and some below the axis, each component must be calculated separately and given its appropriate sign.

▶ Polar coordinates (Section 15.5) define a point, P, in terms of its distance from the origin, O (denoted by the letter r) and the angle (θ) that OP makes with the x-axis.

▶ Areas can be found by taking the definite integral of functions expressed in 'polar' form.

16

Approximate integration

In this chapter you will learn:

▶ *that it is not always possible to integrate exactly*

▶ *how to find good approximations to the value of a definite integral.*

16.1 The need for approximate integration

You cannot always find an area by integration. You may not know the equation of the curve involved; there may not be an equation for the curve; or you may not be able to carry out the integration even if you did know the equation.

In these circumstances, you have to approximate to the area. There are certain practical methods for finding area, such as using squared paper and counting squares, which yield rough approximate results; but there are also methods of calculation by which you can find the area with greater accuracy, though still only approximately. The first of these is the trapezoidal rule.

16.2 The trapezoidal rule

Let the region for which you need to find the area be enclosed by the curve PV (Figure 16.1), the x-axis and the ordinates PA and VG. Divide AG into any number of equal parts, at B, $C, D, \ldots$, each of length h, and draw the corresponding ordinates $PA, QB, RC, \ldots, VG$.

Join $PQ, QR, RS, \ldots, UV$ Let the lengths of the ordinates be $y_0, y_1, y_2, y_3, \ldots, y_n$. Then each of the figures formed by these constructions, such as $APQB$, is a trapezium, and the sum of their areas is

$$\tfrac{1}{2}h\{(y_0 + y_1) + (y_1 + y_2) + (y_2 + y_3) + \ldots + (y_{n-1} + y_n)\}$$

The areas of the trapeziums approximate to the areas of those figures in which the straight line PQ is replaced by the curve PQ, and so on for the others. Consequently the sum of all these approximates to the required area of the whole figure, and the greater the number, the closer in general will be the approximation.

Let the last ordinate GV be y_n. The area is approximately equal to:

$$\tfrac{1}{2}h\{(y_0 + y_1) + (y_1 + y_2) + (y_2 + y_3) + \ldots + (y_{n-1} + y_n)\}$$
$$= \tfrac{1}{2}h\{(y_0 + y_n) + 2(y_1 + y_2 + y_3 + \ldots + y_{n-1})\}$$

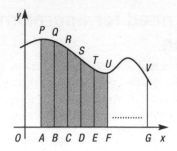

Figure 16.1

Example 16.2.1

Use the trapezoidal rule with ten strips to find an approximation to the integral $\int_0^1 \dfrac{1}{1+x^2}\, dx$

Divide the interval from 0 to 1 into ten equal divisions each of 0.1. Then the lengths of the ordinates will be $y_0, y_1, \ldots, y_{10}$ and these values give the following table, where all the figures are obtained using a calculator and given correct to five decimal places.

First and last	Other ordinates
$y_0 = 1.000\,00$ $y_{10} = 0.500\,00$	$y_1 = 0.990\,10$ $y_2 = 0.961\,54$ $y_3 = 0.917\,43$ $y_4 = 0.862\,07$ $y_5 = 0.800\,00$ $y_6 = 0.735\,29$ $y_7 = 0.671\,14$ $y_8 = 0.609\,76$ $y_9 = 0.552\,49$
Sum = 1.500 00	Sum = 7.099 81

Then, using the trapezoidal rule, the area is approximately:

$$0.1\left(\frac{1.500\,00 + 2 \times 7.099\,81}{2}\right) = 0.784\,98$$

The exact value of the integral is:

$$\int_0^1 \frac{1}{1+x^2}\, dx = \left[\tan^{-1} x\right]_0^1 = \tfrac{1}{4}\pi = 0.785\,40$$

You can see that the approximation is correct to 3 decimal places.

16.3 Simpson's rule for area

Nugget – Thomas Simpson

Simpson's rule (for finding approximations to definite integrals) takes its name from the British mathematician, Thomas Simpson (1710–61). However, Simpson was not the first person to discover this approximation – Johannes Kepler actually beat him to it some 200 years earlier!

Consider again the curve in Section 16.2. If you replace the chords $PQ, QR, RS, \dots, VG$ by arcs of suitable curves, which have equations which you can integrate, the approximation to the area will be closer than the trapezoidal rule approximation.

Assume that the curve joining three consecutive points, such as P, Q, R, is the arc of a parabola. Let the origin for this parabola be B. Then the coordinates of A, B and C are $-h, 0, h$. Then you can find the area of $APRC$ by integration.

Let the equation of the parabola, of which PQR is an arc, be:

$$y = a + bx + cx^2$$

Then, since the equation is satisfied by the coordinates of A, B and C

$AP = a - bh + ch^2 = y_0$		**equation 1**
$BQ = a \qquad\qquad = y_1$		**equation 2**
$CR = a + bh + ch^2 = y_2$		**equation 3**

Therefore $\qquad\qquad a = y_1$

Adding equations **1** and **3**, $y_0 + y_2 = 2(a + ch^2)$

so $\qquad\qquad\qquad 2ch^2 = y_0 + y_2 - 2a = y_0 + y_2 - 2y_1$

so $\qquad\qquad\qquad ch^2 = \frac{1}{2}(y_0 + y_2 - 2y_1)$

Integrating to find the area under the parabola *APRC* gives:

$$\text{Area} = \int_{-h}^{h} (a + bx + cx^2)\, dx$$

$$= \left[ax + \tfrac{1}{2} bx^2 + \tfrac{1}{3} cx^3 \right]_{-h}^{h}$$

$$= 2ah + \tfrac{2}{3} ch^3 = 2h \left(a + \tfrac{1}{3} ch^2 \right)$$

$$= 2h \left\{ y_1 + \tfrac{1}{6} (y_0 + y_2 - 2y_1) \right\}$$

$$= 2h \left(\frac{4y_1 + y_0 + y_2}{6} \right)$$

$$= h \left(\frac{y_0 + 4y_1 + y_2}{3} \right)$$

Similarly, the area under the parabola *RCET* is:

$$h \left(\frac{y_2 + 4y + y_4}{3} \right)$$

Adding these similar expressions gives for the total area:

$$\tfrac{1}{3} h \left\{ (y_0 + 4y_1 + y_2) + (y_2 + 4y_3 + y_4) + \cdots + (y_{n-2} + 4y_{n-1} + y_n) \right\}$$

$$= \tfrac{1}{3} h \left\{ (y_0 + y_n) + 4 (y_1 + y_3 + \cdots) + 2 (y_2 + y_4 + \cdots) \right\}$$

This result is called **Simpson's rule.**

Simpson's rule must always be applied to an even number of intervals, involving an odd number of ordinates. Thus, if there are $2n$ intervals, there will be $2n + 1$ ordinates.

The greater the 'number of strips' which are taken, the greater, in general, will be the accuracy of the approximation to the area.

Example 16.3.1

Use Simpson's rule with ten strips to approximate to $\int_0^1 \dfrac{1}{1+x^2}\,dx$

Divide the interval from 0 to 1 into ten equal divisions each of 0.1. Then the ordinates will be represented by $y_0, y_1, \ldots, y_{10}$ and these values give the following table, where all the figures are obtained using a calculator and given correct to five decimal places.

First and last	Even ordinates	Odd ordinates
$y_0 = 1.000\,00$ $y_{10} = 0.500\,00$	$y_2 = 0.961\,54$ $y_4 = 0.862\,07$ $y_6 = 0.735\,29$ $y_8 = 0.609\,76$ $y_9 = 0.552\,49$	$y_1 = 0.990\,10$ $y_3 = 0.917\,43$ $y_5 = 0.800\,00$ $y_7 = 0.671\,14$
Sum = 1.500 00	Sum = 3.168 66	Sum = 3.931 16

Then, using Simpson's rule, the area is approximately:

$$\text{Area} = 0.1\left(\frac{1.500\,00 + 4 \times 3.168\,66 + 2 \times 3.931\,36}{3}\right) = 0.785\,40$$

The exact value of the integral is

$$\int_0^1 \frac{1}{1+x^2}\,dx = \left[\tan^{-1} x\right]_0^1 = \tfrac{1}{4}\pi = 0.785\,40$$

In fact the approximation is correct to 8 decimal places.

Test yourself

1 The lengths of nine equidistant ordinates of a curve are 8, 10.5, 12.3, 11.6, 12.9, 13.8, 10.2, 8 and 6 m respectively, and the length of the base is 24 m. Find an approximation to the area between the curve and the base.

2 An area is divided into ten equal parts by parallel ordinates 0.2 m apart, the first and last touching the bounding curve. The lengths of the ordinates are 0, 1.24, 2.37, 4.10, 5.28, 4.76, 4.60, 4.36, 2.45, 1.62, 0. Find an approximation to the area.

3 Find an approximation to the area under the curve shown in Figure 16.2, the ordinates being drawn at the points marked 0 to 12, and each division representing one metre.

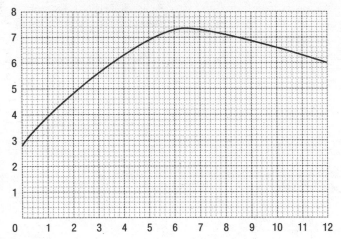

Figure 16.2

4 The lengths of the ordinates of a curve, all in mm, are 2.3, 3.8, 4.4, 6.0, 7.1, 8.3, 8.2, 7.9, 6.2, 5.0, 3.9. Find the area under the curve if each of the ordinates are 1 mm apart.

5 Ordinates at a common distance of 10 m are of lengths, all in metres, 6.5, 9, 13, 18.5, 22, 23, 22, 18.5, 14.5. Find an approximation to the area bounded by the curve, the x-axis and the end ordinates.

6 Show that Simpson's rule with two strips gives the exact value for the area under the graph of $y = x^3$ between $x = a$ and $x = b$.

7 Use the trapezoidal rule with five ordinates to find an approximation to the area under the graph of $y = \sqrt{1 + x^3}$ between $x = 0$ and $x = 1$. Give your result correct to 5 decimal places. Compare your result with the answer to question 8.

8 Use Simpson's rule with five ordinates to find an approximation to the area under the graph of $y = \sqrt{1 + x^3}$ between $x = 0$ and $x = 1$. Give your result correct to 5 decimal places. Compare your result with the answer to question 7.

17

Volumes of revolution

In this chapter you will learn:

- ▶ *what a solid of revolution is*
- ▶ *how to use integration to find a formula for the volume of a solid of revolution.*

17.1 Solids of revolution

The methods of integration which helped you to find areas of plane figures may be extended to the volumes of regular solids.

The solids we will now look at are those which are swept out in space when a curve or area is rotated about some axis. These solids are called **solids of revolution**. For example, if a semicircle is rotated about its diameter it will generate a sphere. Similarly, a rectangle rotated about one side will describe a cylinder.

Nugget – Knitting needle

It is often useful to extend your mental imagery to help grasp the essential features of a mathematical idea, and sometimes it is worth bringing in objects from the everyday world. This is particularly helpful when, as you will do shortly, you are asked to move from two-dimensional images to those in three dimensions.

Imagine what would happen if you were to draw a straight line through the origin and then, in three dimensions, spin this line around the x-axis. The effect would be to create a conical shape lying on its side with the apex of the cone at the origin and the x-axis skewering the cone through the middle. This is essentially what is going on in Figure 17.1b, where the line *OM* represents the straight line.

17.2 Volume of a cone

The method used to find the volumes of solids of revolution can be illustrated by the example of a cone. If a right-angled triangle rotates completely about one of the sides adjacent to the right angle as an axis, the solid generated is a cone.

Take the x-axis as the axis of rotation, the line $y = mx$ as the hypotenuse of the right-angled triangle OAM, and h as the height OM of the cone. See Figure 17.1a. Then the base radius r of the cone is given by:

$$m = \frac{r}{h} \qquad \text{so } r = mh$$

In Figure 17.1b the line OA has been rotated about the x-axis to show the cone.

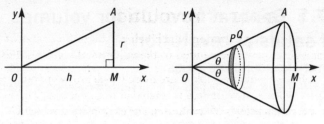

Figure 17.1a   **Figure 17.1b**

Let V be the volume of the cone with vertex O, height h and base radius r. Let P be any point on OA and let its coordinates be (x, y).

Let x be increased by δx so that the corresponding point Q on OA has coordinates $(x + \delta x, y + \delta y)$.

PQ, on rotating, describes a thin disk-shaped slice of the cone, the end faces of which are the circles described by P and Q.

The thickness of the slice is δx.

Its volume lies between the cylinders with volumes $\pi y^2\, \delta x$ and $\pi(y + \delta y)^2\, \delta x$.

If Q is close to P and δx becomes small, then the volume of this slice becomes closer to $\pi y^2\, \delta x$. The element of volume is therefore $\pi y^2\, \delta x$.

The volume of the cone is the limit as $\delta x \to 0$ of the sum of all such elements between the values $x = 0$ and $x = h$, that is:

$$V = \lim_{\delta x \to 0} \sum_{x=0}^{x=h} \pi y^2 \delta x$$

So, using Section 15.2

$$V = \int_0^h \pi y^2 \, dx$$
$$= \pi \int_0^h (mx)^2 \, dx$$
$$= \pi m^2 \left[\tfrac{1}{3} x^3 \right]_0^h$$
$$= \tfrac{1}{3} \pi m^2 h^3$$

But $m = \dfrac{r}{h}$, so $V = \tfrac{1}{3} \pi \dfrac{r^2}{h^2} h^3 = \tfrac{1}{3} \pi r^2 h$

17.3 General formula for volumes of solids of revolution

ROTATION AROUND THE X-AXIS

Let AB (Figure 17.2) be part of a curve with equation $y = f(x)$.

Let AB rotate around the x-axis, generating a solid which is shown in Figure 17.2. Let $OM = a$, $ON = b$. Let P be any point on AB and let its coordinates be (x, y).

Let x be increased by δx so that the corresponding point Q on AB has coordinates $(x + \delta x, y + \delta y)$.

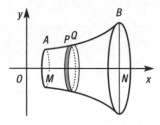

Figure 17.2

PQ, on rotating, describes a thin disk-shaped slice, the end faces of which are the circles described by P and Q.

The thickness of the slab is δx.

Its volume lies between the cylinders with volumes $\pi y^2 \, \delta x$ and $\pi(y + \delta y)^2 \, \delta x$.

If Q is close to P and δx becomes small, then the volume of this slice becomes closer to $\pi y^2 \, \delta x$.

The element of volume is therefore $\pi y^2 \, \delta x$.

The volume of the whole solid is the limit as $\delta x \to 0$ of the sum of all such slices between the limits $x = a$ and $x = b$. Let V be this volume. Therefore:

$$V = \lim_{\delta x \to 0} \sum_{x=a}^{x=b} \pi y^2 \delta x$$

and (using Section 17.2)

$$V = \int_a^b \pi y^2 \, dx.$$

Since $y = f(x)$ you can substitute for y in terms of x and integrate.

Nugget – Stack of coins side by side

Here is another piece of imagery that you may find helpful when thinking about volumes of revolution. Look at Figure 17.2 and focus on the little coin-shaped disc in the middle of the picture marked with the letters *PQ*. This shape is a cylinder of radius *y*. To find its volume, you need to multiply the area of its circular face (πy^2) by its thickness (δx) giving a volume of $\pi y^2 \delta x$.

Now consider a stack of such little discs standing on edge side by side along the horizontal axis. Their combined volume will be the sum of each of the volumes of all these little cylinders.

This explains in simple terms the formula for *V* above.

ROTATION AROUND THE Y-AXIS

Let *AB* (Figure 17.3) be part of a curve with equation $y = f(x)$.

Let *AB* rotate around the y-axis, generating a solid which is shown in Figure 17.3.

Let $UM = a$, $UN = b$. Let *P* be any point on AB and let its coordinates be (x, y).

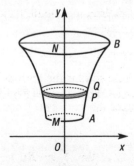

Figure 17.3

Let x be increased by δx so that the corresponding point Q on AB has coordinates $(x + \delta x, y + \delta y)$.

PQ, on rotating, describes a thin disc-shaped slice, the end faces of which are the circles described by P and Q.

The thickness of the slab is δy.

Its volume lies between the cylinders with volumes $\pi x^2 \delta y$ and $\pi(x + \delta x)^2 \delta y$.

If Q is close to P and δx becomes small, then the volume of this slice becomes closer to $\pi y^2 \delta x$.

The element of volume is therefore $\pi x^2 \delta y$.

The volume of the whole solid is the limit as $\delta y \to 0$ of the sum of all such slices between the limits $y = a$ and $y = b$. Let V be this volume. Therefore:

$$V = \lim_{dx \to 0} \sum_{y=a}^{y=b} \pi x^2 \delta y$$
$$= \int_a^b \pi x^2 \, dy$$

Since $y = f(x)$ you can find x in terms of y, substitute in the integral and integrate.

Nugget – Stack of coins one above the other

Let's now extend the coin imagery to support your understanding of volumes of revolution about the y-axis, as shown in Figure 17.3. This time, think of a pile of coins stacked vertically. (In fact, in this particular diagram, the coins will get wider as you move up the pile.) This time the radius of the circular face is x and the thickness is ∂y, giving a volume of $\pi x^2 \partial y$.

The combined volume of a vertical pile of such discs between the limits $y = a$ and $y = b$ will be the sum of each of the volumes of all these little cylinders.

This explains in simple terms the formula for V above.

17.4 Volume of a sphere

Let the equation of the circle in Figure 17.4 be $x^2 + y^2 = a^2$.
The centre is at the origin and its radius $OA = a$. Rotate the
quadrant OAB about the x-axis. The volume described will be a
hemisphere.

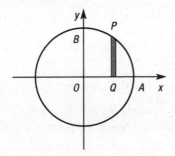

Figure 17.4

Using the formula $V = \int_a^b \pi y^2 dx$. (of Section 17.2), and calling the
volume of the sphere V, you find:

$$V = 2 \times \int_0^a \pi y^2 dx$$
$$= 2\pi \int_0^a (a^2 - x^2) dx$$
$$= 2\pi \left[a^2 x - \tfrac{1}{3} x^3 \right]_0^a$$
$$= 2\pi \left(a^3 - \tfrac{1}{3} a^3 \right)$$
$$= \frac{4}{3} \pi a^3$$

17.5 Examples

Example 17.5.1

The part of the parabola $y^2 = 4x$ between $x = 0$ and $x = 4$ is rotated
about the x-axis. Find the volume generated.

Call the volume V, then

$$V = \int_0^4 \pi y^2 \, dx$$

$$= \pi \int_0^4 4x \, dx$$

$$= \pi \left[2x^2 \right]_0^4 = 32\pi \text{ cubic units}$$

Example 17.5.2

The part of the parabola $y^2 = 4x$ between $x = 0$ and $x = 2$ is rotated about the y-axis. Find the volume generated.

Call the volume V, then

$$V = \int_a^b \pi x^2 \, dy$$

where the limits a and b are y-limits.

Note that when $x = 0$, $y = 0$, and when $x = 2$, $y^2 = 8$, so $y = 2\sqrt{2}$. The limits are therefore 0 and $2\sqrt{2}$.

Then
$$V = \int_0^{2\sqrt{2}} \pi x^2 \, dy$$

$$= \pi \int_0^{2\sqrt{2}} \tfrac{1}{16} y^4 \, dy$$

$$= \tfrac{1}{16} \pi \left[\tfrac{1}{5} y^5 \right]_0^{2\sqrt{2}}$$

$$= \tfrac{1}{80} \pi \times 32 \times 4\sqrt{2}$$

$$= \frac{8\sqrt{2}}{5} \pi \text{ cubic units}$$

Nugget – It boils down to summing volumes of cylinders

The basic method for finding volumes of revolution is always the same, as follows.

The underlying shape you are dealing with is a (very thin) cylinder that could be described as a disc.

To find its volume, you need to follow these four steps:

▶ find its radius (this will be y when the rotation is about the x-axis and x when the rotation is about the y-axis)

▶ find its thickness (this will be δx when the rotation is about the x-axis and δy when the rotation is about the y-axis)

▶ apply the formula for the volume of a cylinder, expressing the result either in terms of just x or just y

▶ integrate the result between the specified limits.

Test yourself

1 Find the volume generated by the arc of the curve $y = x^2$
   a when it rotates round the x-axis between $x = 0$ and $x = 3$
   b when it rotates round the y-axis between $x = 0$ and $x = 2$

2 Find the volume generated when an arc of the graph of $y = x^3$
   a rotates round the x-axis between $x = 0$ and $x = 3$
   b rotates round the y-axis between $x = 0$ and $x = 2$

3 Find the volume of the cone formed by the rotation round the x-axis of that part of the line $2x - y + 1 = 0$, intercepted between the axes.

4 The circle $x^2 + y^2 = 9$ rotates round a diameter which coincides with the x-axis. Find:
   a the volume of the segment between the planes perpendicular to the x-axis at distances of 1 unit and 2 units from the centre, and on the same side of it.
   b the volume of the spherical cap cut off by the plane that is 2 units from the centre.

5 Find the volume generated by the rotation of the ellipse $x^2 + 4y^2 = 16$, about its major axis.

6 Find the volume created by the rotation round the x-axis of the part of the curve $y^2 = 4x$ between the origin and $x = 4$.

7 Find the volume generated by rotating one branch of the hyperbola $x^2 - y^2 = a^2$ about the x-axis, between the limits $x = 0$ and $x = 2a$.

8 Find the volume of the solid generated by the rotation round the y-axis of the part of the curve of $y^2 = x^3$ which is between the origin and $y = 8$.

9 Find the volume of the solid generated by the rotation about the x-axis of the part of the curve of $y = \sin x$, between $x = 0$ and $x = \pi$.

10 Find the volume generated by the rotation round the x-axis of the part of the curve of $y = x(x - 2)$ which lies below the x-axis.

11 If the graph of $xy = 1$ is rotated about the x-axis, find the volume generated by the part of the curve between $x = 1$ and $x = 4$.

12 The parabolas $y^2 = 4x$ and $x^2 = 4y$ intersect and the area included between the curves is rotated round the x-axis. Find the volume of the solid generated.

18

Lengths of curves

In this chapter you will learn:

▶ *how to use integration to find the length of a curve*

▶ *how to find the length of a curve in polar coordinates.*

18.1 Lengths of arcs of curves

Similar methods to those for finding areas of regions, and volumes of solids of revolution, can be used to find the lengths of curves. The process, as before, is: to find an element of length; sum these elements; take the limit of this sum as $\delta x \rightarrow 0$; and then use the result of Section 15.2 to change the limit of the sum to an integral.

To find the length of an arc of a curve, let AB (Figure 18.1) represent a part of the graph of a function $y = f(x)$ between the points A (where $x = a$), and B (where $x = b$).

Let P and Q be two points on the curve, and let PQ be the chord through them. Let P be (x, y), and let Q be $(x + \delta x, y + \delta y)$.

Let s be the length of the arc from A to P, and let $s + \delta s$ be the length from A to Q. Then the length of the arc PQ is δs.

Then by Pythagoras's theorem, the length of the chord PQ is given by

$$PQ = \sqrt{(\delta x)^2 + (\delta y)^2}$$

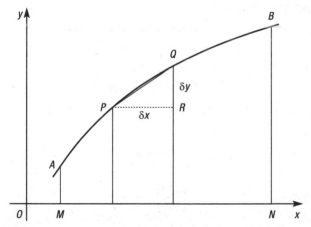

Figure 18.1

If Q is close to P, so that δx is small, the length PQ of the chord is nearly equal to the length δs of the arc. In the limit, as $\delta x \to 0$, the length of the arc is:

$$s = \lim_{\delta x \to 0} \sum_{x=a}^{x=b} \sqrt{(\delta x)^2 + (\delta y)^2}$$

$$= \lim_{\delta x \to 0} \sum_{x=a}^{x=b} \sqrt{1 + \left(\frac{\delta y}{\delta x}\right)^2}\, \delta x$$

$$= \int_{\theta_1}^{b} \sqrt{1 + \left(\frac{dy}{dx}\right)^2}\, dx$$

If it is more convenient to integrate with respect to y, then:

$$s = \int_{c}^{d} \sqrt{1 + \left(\frac{dx}{dy}\right)^2}\, \delta y$$

where c and d are the limits of y.

Example 18.1.1

Find the length of the circumference of the circle $x^2 + y^2 = a^2$.

Since
$$x^2 + y^2 = a^2$$
$$y = \sqrt{a^2 - x^2} = (a^2 - x^2)^{\frac{1}{2}}$$
$$\frac{dy}{dx} = \tfrac{1}{2}(a^2 - x^2)^{-\frac{1}{2}} \times -2x$$
$$= \frac{-x}{\sqrt{a^2 - x^2}}$$

Therefore
$$\left(\frac{dy}{dx}\right)^2 = \frac{x^2}{a^2 - x^2}$$

and
$$1 + \left(\frac{dy}{dx}\right)^2 = 1 + \frac{x^2}{a^2 - x^2} = \frac{a^2}{a^2 - x^2}$$

Considering the length of a quadrant, the limits are 0 and a.
Using the formula above:

$$\text{length} = \int_0^a \sqrt{1+\left(\frac{dy}{dx}\right)^2}\, dx$$

$$= \int_0^a \sqrt{\frac{a^2}{a^2 - x^2}}\, dx$$

$$= a \int_0^a \frac{dx}{\sqrt{a^2 - x^2}}$$

$$= a \left[\sin^{-1}\frac{x}{a} \right]_0^a$$

$$= a \left(\tfrac{1}{2}\pi - 0 \right) = \tfrac{1}{2}\pi a$$

The circumference of the circle is therefore $4 \times \tfrac{1}{2}\pi a = 2\pi a$

Example 18.1.2

Find the length of the arc of the parabola $y = \tfrac{1}{4}x^2$ from the vertex to the point where $x = 2$.

if $\qquad\qquad y = \tfrac{1}{4}x^2, \text{then } \dfrac{dy}{dx} = \tfrac{1}{2}x$

Figure 18.2 shows a sketch of the curve where OQ is the part of the curve whose length you need. The limits for x are 0 and 2.

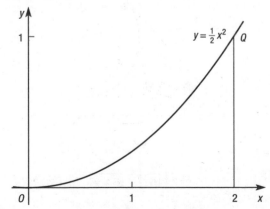

Figure 18.2

296

Therefore
$$\text{length} = \int_0^2 \sqrt{1 + \left(\frac{dy}{dx}\right)^2}\, dx$$
$$= \int_0^2 \sqrt{1 + \tfrac{1}{4}x^2}\, dx$$
$$= \tfrac{1}{2}\int_0^2 \sqrt{4 + x^2}\, dx$$
$$= \tfrac{1}{2}\left[\tfrac{1}{2}x\sqrt{4 + x^2} + \tfrac{4}{2}\ln\left| \frac{x + \sqrt{x^2 + 4}}{2}\right| \right]_0^2$$
$$= \tfrac{1}{2}\left[\tfrac{1}{2}2\sqrt{8} + 2\ln(1 + \sqrt{2}) - 2\ln 1 \right]$$
$$= \sqrt{2} + \ln(1 + \sqrt{2})$$

18.2 Length in polar coordinates

The general method is similar to the one used for rectangular coordinates. In Figure 18.3 let AB be part of a curve for which the polar equation is known.

Let the angles made by OA and OB with OX be θ_1 and θ_2.

Let P and Q be two points on the curve, and let PQ be the chord through them.

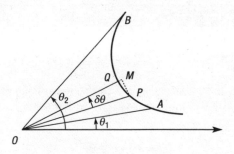

Figure 18.3

Let P be the point (r, θ) and Q be $(r + \delta r, \theta + \delta\theta)$. Let PQ be the chord joining P to Q.

Let s be the length of the arc from A to P, and let $s + \delta s$ be the length from A to Q. Then the length of the arc PQ is δs.

Then $QM = \delta r$ and the arc $PQ = \delta s$. With the construction shown in Figure 18.3, $PM = r\delta\theta$. Then by Pythagoras's theorem, the length of the chord PQ is given by:

$$PQ^2 = (r\delta\theta)^2 + (\delta r)^2$$

If Q is close to P, so that $\delta\theta$ is small, the length of the chord is nearly equal to the length of the arc.

In the limit, as $\delta\theta \to 0$, the length of the arc is:

$$\text{length} = \lim_{\delta\theta \to 0} \sum_{x=a}^{x=b} \sqrt{(r\delta\theta)^2 + (\delta r)^2}$$

$$= \lim_{\delta\theta \to 0} \sum_{x=a}^{x=b} \sqrt{r^2 + \left(\frac{\delta r}{\delta\theta}\right)^2}\, \delta\theta$$

$$= \int_{\theta_1}^{\theta_2} \sqrt{r^2 + \left(\frac{dr}{d\theta}\right)^2}\, d\theta$$

If it is more convenient to integrate with respect to r, then:

$$\text{length} = \int_{r_1}^{r_2} \sqrt{1 + r^2 \left(\frac{d\theta}{dr}\right)^2}\, dr$$

Example 18.2.1

Find the complete length of the cardioid with equation $r = a(1 + \cos\theta)$.

Section 15.6 (the construction of a complete cardioid) involved a complete rotation of the radius vector, so that θ increases from 0 to 2π.

Since
$$r = a(1 + \cos\theta)$$

$$\frac{dr}{d\theta} = -a\sin\theta$$

Since
$$s = \int_{\theta_1}^{\theta_2} \sqrt{r^2 + \left(\frac{dr}{d\theta}\right)^2}\, d\theta$$

you have, using the symmetry of the curve:

$$\text{length} = 2 \int_0^\pi \sqrt{a^2(1+\cos\theta)^2 + (-a\sin\theta)^2} \, d\theta$$

$$= 2a \int_0^\pi \sqrt{1 + 2\cos\theta + \cos^2\theta + \sin^2\theta} \, d\theta$$

$$= 2a \int_0^\pi \sqrt{2 + 2\cos\theta} \, d\theta$$

$$= 2a \int_0^\pi 2\cos\tfrac{1}{2}\theta \, d\theta$$

$$= 4a \times \left[2\sin\tfrac{1}{2}\theta \right]_0^\pi$$

$$= 8a$$

An important point arises if you try to avoid the use of symmetry by saying

$$\text{length} = \int_0^{2\pi} \sqrt{a^2(1+\cos\theta)^2 + (-a\sin\theta)^2} \, d\theta$$

When you carry out the integration, you appear to get the result:

$$\text{length} = \int_0^{2\pi} \sqrt{a^2(1+\cos\theta)^2 + (-a\sin\theta)^2} \, d\theta$$

$$= a \int_0^{2\pi} \sqrt{1 + 2\cos\theta + \cos^2\theta + \sin^2\theta} \, d\theta$$

$$= a \int_0^{2\pi} \sqrt{2 + 2\cos\theta} \, d\theta$$

$$= a \int_0^{2\pi} 2\cos\tfrac{1}{2}\theta \, d\theta$$

$$= 2a \times \left[2\sin\tfrac{1}{2}\theta \right]_0^{2\pi}$$

$$= 0$$

What has happened? Where is the error?

You must remember that square roots must always be positive, so, when you take the square root of the expression $2 + 2\cos\theta$ you must think which of the two possibilities, $2\cos\tfrac{1}{2}\theta$ and $-2\cos\tfrac{1}{2}\theta$, is positive. If you do this, you find that in the interval from 0 to π, $2\cos\tfrac{1}{2}\theta$ is positive, but in the interval from π to 2π, $-2\cos\tfrac{1}{2}\theta$ is positive.

The fourth line of the argument above should therefore be:

$$a \int_0^\pi 2\cos\tfrac{1}{2}\theta \, d\theta + a \int_\pi^{2\pi} \left(-2\cos\tfrac{1}{2}\theta \right) d\theta$$

Evaluating this expression then gives:

$$\text{length} = a \int_0^\pi 2\cos\tfrac{1}{2}\theta \, d\theta + a \int_\pi^{2\pi} \left(-2\cos\tfrac{1}{2}\theta\right) d\theta$$
$$= 2a \times \left[2\sin\tfrac{1}{2}\theta\right]_0^\pi + 2a \times \left[2\sin\tfrac{1}{2}\theta\right]_\pi^{2\pi}$$
$$= 4a + 4a$$
$$= 8a$$

Nugget – Let's hear it for the calculus!

Throughout much of the history of mathematics, most of the great minds believed that, although straight lines had definite lengths, curves did not. It was believed impossible to find the length of 'an irregular arc'. In fact it was the invention of calculus that enabled mathematicians to solve this problem by taking ever better approximations to the answer.

Test yourself

1 Find the length of the arc of the parabola $y = \tfrac{1}{2}x^2$ between the origin and the ordinate $x = 2$.

2 Find the length of the arc of the parabola $y^2 = 4x$ from $x = 0$ to $x = 4$.

3 Find the length of the arc of the curve $y^2 = x^3$ from $x = 0$ to $x = 5$.

4 Find the length of the arc of the catenary $y = \cosh x$ from the vertex to the point where $x = 1$.

5 Find the length of the arc of the curve $y = \ln x$ between the points where $x = 1$ and $x = 2$. (To carry out the integration, rationalize the numerator.)

6 Find the length of the part of the curve of $y = \ln \sec x$ between the values $x = 0$ and $x = \tfrac{1}{3}\pi$.

7 Find the length of the circumference of the circle with equation $r = 2a\cos\theta$

8 Find the whole length of the curve $r = a\sin^3\tfrac{1}{3}\theta$.

19

Taylor's and Maclaurin's series

In this chapter you will learn:

▶ *how to find polynomial approximations to functions*

▶ *the meanings of 'convergent' and 'divergent' as applied to series*

▶ *the methods of Taylor and Maclaurin.*

19.1 Infinite series

When studying algebra you learned about certain series such as, for example, the geometric progression (or series), the arithmetic series and the binomial series.

For the geometric series, it is likely that you will have considered the important problem of the sum of the series, when the number of terms is increased without limit, that is, it becomes infinite. Two cases arise:

1 When the common ratio r is numerically greater than unity, as the number of terms increases, the terms themselves get bigger and so does their sum. If the number of terms becomes infinite, their sum also becomes infinite, that is, if S_n is the sum of n terms, then, when $n \to \infty$, $S_n \to \infty$.

2 If, however, the common ratio r is less than unity, the terms continually decrease, and the question of what happens to S_n as n approaches infinity is a matter for investigation.

In this case, you can prove that when $n \to \infty$, S_n approaches a finite limit.

19.2 Convergent and divergent series

In general, when considering any kind of series, the problem you need to investigate is whether:

1 S_n approaches a finite limit when $n \to \infty$, or

2 S_n approaches infinity when $n \to \infty$

A series of the first kind is said to be **convergent**; a series of the second kind is called **divergent**. (There is also a third type of series called **oscillating**, but this is not considered in this chapter.)

For theoretical and practical purposes, it is very important to know whether a given series is convergent or divergent. There is no universal method of determining this, but there are various tests which can be applied for certain kinds of series. A consideration of these tests is, however, beyond the scope of this book.

In this brief treatment of infinite series using calculus, the series considered will be assumed (without proof) to be convergent.

19.3 Taylor's expansion

You can, using the binomial expansion, expand the function $(x + a)^n$ as a series of descending powers of x and ascending powers of a. You can also expand many other functions, using a variety of methods. However, in this chapter, a general method of expanding functions in series is investigated.

Briefly, you will see that $f(x + h)$ can, in general, be expanded in a series of ascending powers of h. Such an expansion is not possible for all functions, and there are limitations to the application of the method.

Here is a statement of the result known as Taylor's expansion. The proof follows.

TAYLOR'S EXPANSION

$$f(x + h) = f(x) + \frac{h}{1!} f'(x) + \frac{h^2}{2!} f''(x) + \frac{h^3}{3!} f'''(x) + \cdots + \frac{h^n}{n!} f^n(x) + \cdots$$

The following assumptions will be made.

1 Any function which will be considered is capable of being expanded in this form.

2 The series is convergent (subject in some cases to certain conditions).

3 The successive derivatives, $f'(x), f''(x), f'''(x), \ldots, f^n(x)$, all exist.

In accordance with assumption **1**, assume that $f(x + h)$ can be expanded in ascending powers of h as:

$$f(x + h) = A_0 + A_1 h + A_2 h^2 + A_3 h^3 + \cdots \qquad \textbf{equation 1}$$

(where the coefficients A_0, A_1, A_2, are functions of x but do not contain h)

Since this is to be true for all values of h, let $h = 0$

Then on substitution in equation **1**, you find $A_0 = f(x)$

Since the series in equation **1** is an identity, you can assume that if you differentiate both sides with respect to h, keeping x constant, the result in each case will be another identity.

Repeating the process, you find:

(i) $f'(x + h) = A_1 + 2A_2 h + 3A_3 h^2 + 4A_4 h^3 + \cdots$
   since $f'(x) = 0$ when you differentiate with respect to h, as x is kept constant.

Similarly:

(ii) $f''(x + h) = 2A_2 + 3 \times 2 \times A_3 h + 4 \times 3 \times A_4 h^2 + \cdots$

(iii) $f'''(x + h) = 3 \times 2 \times A_3 + 4 \times 3 \times 2 \times A_4 h +$
   and so on for higher derivatives.

In all of these results put $h = 0$. Then:

(from (i)) $f'(x) = A_1$
(from (ii)) $f''(x) = 2A_2$
(from (iii)) $f'''(x) = 3 \times 2 \times A_3$
                  $\vdots$

which leads to $A_1 = f'(x)$

$$A_2 = \frac{f''(x)}{2!}$$

$$A_3 = \frac{f'''(x)}{3!}$$

$$\vdots$$

and so on $A_n = \frac{f^n(x)}{n!}$

Substituting for these in equation 1 you obtain Taylor's expansion:

$$f(x+h) = f(x) + \frac{h}{1!}f'(x) + \frac{h^2}{2!}f''(x) + \frac{h^3}{3!}f'''(x) + \cdots + \frac{h^n}{n!}f^n(x) + \cdots$$

Example 19.3.1

Use Taylor's expansion to expand $(x+h)^n$

Let $f(x+h) = (x+h)^n$

$$= f(x) + \frac{h}{1!}f'(x) + \frac{h^2}{2!}f''(x) + \frac{h^3}{3!}f'''(x) + \cdots$$

When $h = 0$

$$f(x) = x^n$$
$$f'(x) = nx^{n-1}$$
$$f''(x) = n(n-1)x^{n-2}$$
$$f'''(x) = n(n-1)(n-2)x^{n-3}$$

Substituting into the expression for Taylor's expansion

$$(x+h)^n = x^n + hnx^{n-1} + \frac{h^2}{2!}n(n-1)x^{n-2}$$
$$+ \frac{h^3}{3!}n(n-1)(n-2)x^{n-3} + \cdots$$

and re-arranging

$$(x+h)^n = x^n + nx^{n-1}h + \frac{n(n-1)}{2!}x^{n-2}h^2$$
$$+ \frac{n(n-1)(n-2)}{3!}x^{n-3}h^3 + \cdots$$

19.4 Maclaurin's series

Nugget – Colin Maclaurin

Colin Maclaurin (1698–1746) was born in Argyll in Scotland and, at 11 years of age, he entered the University of Glasgow. At 19 he was

Maclaurin's series is a special case of Taylor's series. You obtain it first by putting $x = 0$, and then, for convenience, replacing h by x.

This is possible since Taylor's expansion is true for all values of x and h.

Therefore, let $x = 0$, and replace x by h.

Then Taylor's expansion becomes:

$$f(x) = f(0) + \frac{x}{1!}f'(0) + \frac{x^2}{2!}f''(0) + \frac{x^3}{3!}f'''(0) + \cdots + \frac{x^n}{n!}f^n(0) + \cdots$$

In this form $f^n(0)$ means that in the nth derivative of $f(x)$, x is replaced by 0.

Example 19.4.1

Expand $\ln(1 + x)$ as a series in ascending power of x.

Let $\qquad f(x) = \ln(1 + x) \qquad$ then $f(0) = \ln 1 = 0$

$\qquad f'(x) = \dfrac{1}{1 + x} \qquad$ then $f'(0) = \dfrac{1}{1} = 1$

$\qquad f''(x) = \dfrac{-1}{(1 + x)^2} \qquad$ then $f''(0) = \dfrac{-1}{1} = -1$

$\qquad f'''(x) = \dfrac{1 \times 2}{(1 + x)^3} \qquad$ then $f'''(0) = \dfrac{1 \times 2}{(1 + x)^3} = 1 \times 2$

$$\vdots$$

$$f^n(x) = (-1)^{n+1} \frac{(n-1)!}{(1+x)^n}, \text{ then } f^n(0) = (-1)^{n-1}(n-1)!$$

Substituting these values in Maclaurin's series

$$f(x) = \ln(1+x) = f(0) + \frac{x}{1!}f'(0) + \frac{x^2}{2!}f''(0) + \cdots$$

and

$$\ln(1+x) = x - \frac{x^2}{2} + \frac{x^3}{3} - \frac{x^4}{4} + \cdots + (-1)^n \frac{x^n}{n} + \cdots$$

Remember that throughout this work the logarithms have been natural logarithms to the base e. Consequently the series for $\ln(1+x)$ may be used to calculate logarithms to the base e. From these the logarithms to any other base, such as 10, can be obtained.

It is interesting to look at the graphs of $\ln(1+x)$, together with the graphs of successive Maclaurin polynomial approximations $x, x - \frac{x^2}{2}, x - \frac{x^2}{2} + \frac{x^3}{3}$ and $x - \frac{x^2}{2} + \frac{x^3}{3} - \frac{x^4}{4}$. These graphs are shown in Figure 19.1.

Graph 1 is $y = x$

Graph 2 is $y = x - \frac{x^2}{2}$

Graph 3 is $y = x - \frac{x^2}{2} + \frac{x^3}{3}$

Graph 4 is $y = x - \frac{x^2}{2} + \frac{x^3}{3} - \frac{x^4}{4}$

You can see from Figure 19.1 that the approximations near to $x = 0$ are very good, but as you get further from $x = 0$ the approximations are less good. In particular at $x = -1$, the function $\ln(1+x)$ does not exist as there is no value for $\ln 0$; this means that the series cannot converge for $x = -1$. In fact, this series converges only for $-1 < x \le 1$.

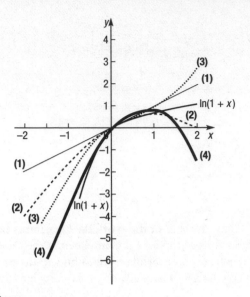

Figure 19.1

Example 19.4.2

Expand $\sin x$ in a series involving ascending powers of x.

Let $\quad f(x) = \sin x \qquad\qquad$ then $f(0) = 0$

$\quad f'(x) = \cos x = \sin\left(x + \tfrac{1}{2}\pi\right) \quad$ then $f'(0) = 1$

$\quad f''(x) = -\sin x = \sin\left(x + \tfrac{2}{2}\pi\right)$ then $f''(0) = 0$

$\quad f'''(x) - \cos x = \sin\left(x + \tfrac{3}{2}\pi\right) \quad$ then $f'''(0) = -1$

$\qquad\vdots \qquad\qquad\qquad\qquad\qquad\qquad \vdots$

$\quad f^n(x) = \sin\left(x + \dfrac{n}{2}\pi\right) \qquad\qquad$ then $f^n(0) = \sin\left(\dfrac{n\pi}{2}\right)$

Substituting these values in Maclaurin's series, you find:

$$\sin x = x - \frac{x^3}{3!} + \frac{x^5}{5!} - \frac{x^7}{7!} + \dots + \frac{x^n \sin\dfrac{n\pi}{2}}{n!} + \dots$$

In this series x is measured in radians.

If you put $x = 1$, you can calculate sin 1 by taking sufficient terms of the series. You can see that the terms decrease rather rapidly.

Note that the series contains only odd powers of x, that is, it is an **odd function**. The series for cos x turns out to contain only even powers of x, that is, it is an **even function**.

Example 19.4.3

Expand e^x in a series involving ascending powers of x.

Let
$$f(x) = e^x \qquad \text{then } f(0) = 1$$
$$f'(x) = e^x \qquad \text{then } f'(0) = 1$$
$$f''(x) = e^x \qquad \text{then } f''(0) = 1$$

Substituting these values in Maclaurin's series, you find:

$$e^x = 1 + \frac{x}{1!} + \frac{x^2}{2!} + \frac{x^3}{3!} + \frac{x^4}{4!} + \cdots$$

19.5 Expansion by the differentiation and integration of known series

Example 19.5.1 illustrates the method.

Example 19.5.1

Using division, you find that:

$$\frac{1}{1+x^2} = 1 - x^2 + x^4 - x^6 + \cdots$$

It may be proved that when a function is represented by a series, and the function and the series are integrated throughout, the results are equal.

Therefore $\displaystyle\int \frac{1}{1+x^2}\, dx = \int dx - \int x^2\, dx + \int x^4\, dx - \cdots$

Therefore $\qquad \tan^{-1} x = x - \dfrac{x^3}{3} + \dfrac{x^5}{5} - \dfrac{x^7}{7} + \cdots$

This series is called **Gregory's series**. It is convergent and can be used to calculate the value of π.

Thus, in the series let $x = 1$

Then
$$\tan^{-1} 1 = \tfrac{1}{4}\pi$$

Substituting in Gregory's series

$$\frac{1}{4}\pi = 1 - \frac{1}{3} + \frac{1}{5} - \frac{1}{7} + \frac{1}{9} - \cdots$$

Hence, by taking sufficient terms, you can find the value of π to any required degree of accuracy. The series converges slowly, however. Consequently other series which converge more rapidly are generally used for calculating π.

Test yourself

Expand the following functions in ascending powers of x, giving the first three non-zero terms.

1 $\sin(a + x)$	**2** $\cos(a + x)$	**3** e^{x+h}
4 $\ln(1 + \sin x)$	**5** $\cos x$	**6** $\tan x$
7 $\ln(1 + e^x)$	**8** ax	**9** e-kx
10 $e^{\sin x}$	**11** $\sec x$	**12** $\ln(\sec x)$
13 $\sin^{-1} x$	**14** $\ln(1 - x)$	**15** $\sinh x$
16 $e^x \sin x$	**17** $\tanh x$	

20

Differential equations

In this chapter you will learn:

- ▶ *the meaning of differential equations*
- ▶ *how to solve differential equations of various types.*

20.1 Introduction and definitions

Equations which contain an independent variable, a dependent variable and at least one of their derivatives are called **differential equations**.

Differential equations are of great importance in physics, all kinds of engineering, and other applications of mathematics. It is not possible in this book to give more than a brief introduction to what is a big subject, however, the elementary forms of differential equations dealt with in this chapter may prove valuable to many students.

Examples of differential equations have already appeared in this book (as, for example, questions 49 to 54 in Exercise 11.1).

Again (as illustrated in Section 11.2)

if $$\frac{dy}{dx} = 2x$$ equation 1

or $$dy = 2x\,dx$$ equation 2

you obtain by integration the relation

$$y = x^2 + c$$ equation 3

Equations 1 and 2 are differential equations, and equation 3 is their solution. Thus a differential equation is solved when, by integration, you find the relation between the two variables, in this case, x and y.

This process involves the introduction of an undetermined constant. Thus the solution in equation 3 is the general equation, or the relation between y and x for the whole family of curves represented in Figure 11.1.

Nugget – What it means to 'solve' a differential equation

In conventional algebra, solving an equation is normally taken to mean finding one or more numerical values that make the equation true.

Here are two examples:

1 The unique solution to $3x - 12 = 0$ is $x = 4$

2 The two solutions to $x^2 - x - 2 = 0$ are $x = 2$ and $x = -1$.

Solutions to differential equations look very different to this. As the name suggests, what you start with is an equation involving differentials ($\frac{dy}{dx}$ and such like). The aim in solving them is to end up with an equation that does not involve any differentials.

FORMATION OF DIFFERENTIAL EQUATIONS

Differential equations arise or may be derived in a variety of ways.

For example, it is shown in mechanics that if s is the displacement of a body at time t moving with uniform acceleration, a, then:

$$\frac{d^2s}{dt^2} = a \qquad \text{equation 4}$$

By integration
$$\frac{ds}{dt} = at + c_1 \qquad \text{equation 5}$$

Integrating again
$$s = \tfrac{1}{2}at^2 + c_1 t + c_2 \qquad \text{equation 6}$$

Of these equations, 4 contains a second derivative, 5 the first derivative, while 6 is the general solution of equations 4 and 5. Differential equations may also be formed by direct differentiation.

Thus, let
$$y = x^3 + 7x^2 + 3x + 7 \qquad \text{equation 7}$$

Then
$$\frac{dy}{dx} = 3x^2 + 14x + 3 \qquad \text{equation 8}$$

$$\frac{d^2y}{dx^2} = 6x + 14 \qquad \text{equation 9}$$

$$\frac{d^3y}{dx^3} = 6 \qquad \text{equation 10}$$

Equation 7 is called the **complete primitive** of equation 10.

KINDS OF DIFFERENTIAL EQUATIONS

Two main types of differential equations are:

▶ ordinary differential equations, with only one independent variable

▶ partial differential equations, with more than one independent variable.

This chapter is concerned with ordinary differential equations only.

ORDER

Differential equations of both types are classified according to the highest derivative which occurs in them. Thus of the differential equations 8, 9 and 10 above:

▶ equation 8 is of the first order, having only the first derivative

▶ equation 9 is of the second order

▶ equation 10 is of the third order.

DEGREE

The degree of a differential equation is the highest power of the highest derivative which the equation contains after it has been simplified by clearing radicals and fractions. This is of the second order and third degree:

$$\left(\frac{d^2y}{dx^2}\right)^3 + 3\frac{dy}{dx} = 0$$

This (see Section 18.1) is of the first order and second degree:

$$ds = \sqrt{1 + \left(\frac{dy}{dx}\right)^2}\, dx$$

SOLUTIONS OF A DIFFERENTIAL EQUATION

A solution that is **complete** or **general** must contain a number of arbitrary constants which is equal to the order of the equation. Thus equation 6 above contains two arbitrary constants and is the solution of equation 4, an equation of the second order.

Solutions that are obtained by assigning particular values to the constants (as in Exercise 11.1, question 54), are called **particular solutions**. This chapter will be concerned only with equations of the first order and first degree.

DIFFERENTIAL EQUATIONS OF THE FIRST ORDER AND FIRST DEGREE

Since solutions of differential equations involve integration, it is not possible to formulate rules (as you could with differentiation) that will apply to every type of equation. There are some equations it is not possible to solve. But a large number of equations (including many of practical importance) can be classified into various types, for which solutions can be found by established methods.

20.2 Type I: one variable absent

These may come in one of two forms.

WHEN Y IS ABSENT

The general form is $\dfrac{dy}{dx} = f(x)$ or $dy = f(x)dx$

and the solution is $y = \int f(x)\,dx$

This requires ordinary integration for its solution.

Example 20.2.1

Solve the equation $\dfrac{dy}{dx} = x^4 + \sin x$

$$y = \int (x^4 + \sin x)dx$$
$$= \tfrac{1}{5}x^5 - \cos x + c$$

WHEN X IS ABSENT

The general form is $\dfrac{dy}{dx} = f(y)$ or $dy = f(y)\,dx$

You can rewrite this in the form $\dfrac{dx}{dy} = \dfrac{1}{f(y)}$ or $dx = \dfrac{dy}{f(y)}$

giving
$$x = \int \frac{dy}{f(y)}$$

You then integrate to find the solution.

Example 20.2.2

Solve the equation $\dfrac{dy}{dx} = \tan y$

Hence
$$\frac{dx}{dy} = \frac{1}{\tan y} = \frac{\cos y}{\sin y}$$

so
$$x = \int \frac{\cos y}{\sin y} dy$$

and
$$x = \ln|\sin y| + c$$

20.3 Type II: variables separable

If you can rearrange the terms of the equation in two groups, each containing only one variable, the variables are said to be **separable**. Then the equation takes the form:

$$f(y)\frac{dy}{dx} + F(x) = 0 \text{ or } f(y)dy + F(x)dx = 0$$

in which $F(x)$ is a function of x only, and $f(y)$ a function of y only. The general solution then is

$$\int f(y)dy + \int F(x)dx = c$$

Example 20.3.1

Solve the differential equation $x\,dy + y\,dx = 0$

To separate the variables divide throughout by xy

Then
$$\frac{dy}{y} + \frac{dx}{x} = 0$$

Therefore
$$\int \frac{dy}{y} + \int \frac{dx}{x} = 0$$
$$\ln|y| + \ln|x| = c_1$$

Write the constant c_1 in the form $c_1 = \ln c$

$$\ln|y| + \ln|x| = \ln c$$

giving
$$xy = c$$

The factor $\frac{1}{xy}$ used to multiply throughout to separate the variables is called **an integrating factor**.

Example 20.3.2

Solve the equation $(1+x)y + (1-y)x\dfrac{dy}{dx} = 0$

Multiplying throughout by $\dfrac{1}{xy}$

$$\frac{1+x}{x}dx + \frac{1-y}{y}dy = 0$$

or

$$\left(\frac{1}{x}+1\right)dx + \left(\frac{1}{y}-1\right)dy = 0$$

$$\int\left(\frac{1}{x}+1\right)dx + \int\left(\frac{1}{y}-1\right)dy = 0$$

$$\ln|x| + x + \ln|y| - y = c$$

$$\ln|xy| + (x-y) = c$$

Nugget – Using an integrating factor in four easy steps

The idea of using an *integrating factor* to separate the variables can be summarized into the following four steps:

1 Work out the integrating factor.

2 Multiply the differential equation through by the integrating factor.

3 Rewrite the left-hand side of the equation so that the variables are matched up.

4 Integrate and solve.

Exercise 20.1

Solve the following differential equations.

1 $\dfrac{dy}{dx} + \dfrac{k}{x^2} = 0$

2 $\dfrac{dy}{dx} = \dfrac{y}{a}$

3 $\dfrac{dy}{dx} = \dfrac{y}{x}$

4 $(1+y)dx - (1-x)dy = 0$

5 $(x+1)dy - y\,dx = 0$

6 $\sin x \cos y\,dx = \sin y \cos x\,dy$

7 $xy\dfrac{dy}{dx} = \dfrac{y^2+1}{x^2+1}$

8 $2y\,dx = x(y-1)dy$

9 $y^2 + \sin 2x\dfrac{dy}{dx} = 1$

10 $\dfrac{dy}{dx} = x^2 y$

11 $x\sqrt{y^2-1}\,dx - y\sqrt{x^2-1}\,dy = 0$

12 $\dfrac{1+x^2}{1+y} = xy\dfrac{dy}{dx}$

13 $\dfrac{dy}{dx} = 2xy$

14 The slope of a family of curves is $-\dfrac{y}{x}$. What is the equation of the set?

20.4 Type III: linear equations

An equation of the form:

$$\frac{dy}{dx} + Py = Q$$

where P and Q are constants, or functions of x only, is called a **linear differential equation**.

This is because y and its derivatives are of the first degree.

If you multiply such an equation throughout by the **integrating factor** $e^{\int P\,dx}$, you obtain an equation which you can solve.

When you multiply by this factor, the equation becomes:

$$e^{\int P\,dx}\left(\frac{dy}{dx} + Py\right) = Qe^{\int P\,dx}$$

If you differentiate $ye^{\int P\,dx}$ you get $e^{\int P\,dx}\dfrac{dy}{dx}+e^{\int P\,dx}Py$, which is the left-hand side of the equation. Therefore the solution is:

$$ye^{\int P\,dx}=\int\left(Qe^{\int P\,dx}\right)dx \qquad\qquad \text{equation 11}$$

The procedure in solving this type of differential equation is to begin by finding the integral $e^{\int P\,dx}$, then substitute in equation **11**. The method is shown in the following examples.

Example 20.4.1

Solve the equation $(1-x^2)\dfrac{dy}{dx}-xy=1$

First transform this equation to the general form $\dfrac{dy}{dx}+Py=Q$ to get

$$\frac{dy}{dx}-\frac{x}{1-x^2}y=\frac{1}{1-x^2}$$

Now find the integrating factor $e^{\int P\,dx}$, where, in this case, $P(x)=-\dfrac{x}{1-x^2}$

Then
$$\int P\,dx=-\int\frac{x}{1-x^2}\,dx$$
$$=\tfrac{1}{2}\ln\left|1-x^2\right|$$
$$=\ln\sqrt{1-x^2}$$

Therefore the integrating factor is $e^{\ln\sqrt{1-x^2}}=\sqrt{1-x^2}$

Using the form in equation **11**, you find that:

$$y\sqrt{1-x^2}=\int\frac{1}{1-x^2}\times\sqrt{1-x^2}\,dx$$
$$=\int\frac{dx}{\sqrt{1-x^2}}$$
$$=\sin^{-1}x+c$$

Therefore the solution is:

$$y\sqrt{1-x^2} = \sin^{-1}x + c$$

Example 20.4.2

Solve the equation $\cos x\dfrac{dy}{dx} + y\sin x = 1$

Dividing by cos x

$$\frac{dy}{dx} + y\tan x = \sec x$$

Comparing with equation **11**, $P(x) = \tan x$ so:

$$\begin{aligned} \int P\,dx &= \int \tan x\,dx \\ &= -\ln|\cos x| \\ &= \ln|\sec x| \end{aligned}$$

Using equation **11** and substituting:

$$\begin{aligned} y\sec x &= \int \sec x \times \sec x\,dx \\ &= \int \sec^2 x\,dx \\ &= \tan x + c \end{aligned}$$

Therefore the solution is:

$$y\sec x = \tan x + c$$
or
$$y = \sin x + c\cos x$$

Example 20.4.3

Solve the equation $\dfrac{dy}{dx} + 2xy = 1 + 2x^2$

Comparing with equation **11**, $P(x) = 2x$, so $\int P\,dx = \int 2x\,dx = x^2$

Therefore the integrating factor is e^{x^2}

Substituting into equation **11**:

$$\begin{aligned} ye^{x^2} &= \int (1 + 2x^2)e^{x^2}\,dx \\ &= \int (e^{x^2} + 2x^2 e^{x^2})\,dx \\ &= xe^{x^2} + c \end{aligned}$$

Therefore the solution is:

$$ye^{x^2} = xe^{x^2} + c$$

or $$y = x + ce^{-x^2}$$

Exercise 20.2

Solve the following differential equations.

1 $\dfrac{dy}{dx} - 2xy = 2x$

2 $x\dfrac{dy}{dx} + x + y = 0$

3 $\dfrac{dy}{dx} = y - x$

4 $\dfrac{dy}{dx} + xy = x$

5 $\dfrac{dy}{dx} + ay = e^x$

6 $\dfrac{dy}{dx} + y\tan x = 1$

7 $\dfrac{dy}{dx} - \dfrac{ay}{x} = \dfrac{x+1}{x}$

8 $\tan x \dfrac{dy}{dx} = 1 + y$

9 $e^x dy = (1 - ye^x)dx$

10 $xdy - ay\,dx = (x + 1)dx$

11 $\cos^2 x \dfrac{dy}{dx} + y = \tan x$

12 $x^2 \dfrac{dy}{dx} + xy + 1 = 0$

20.5 Type IV: linear differential equations with constant coefficients

An important special case of linear differential equations is one where the coefficients are constants. Here are some examples.

$$\frac{dy}{dx} + 2y = 0 \qquad \qquad \textbf{equation 12}$$

$$\frac{d^2y}{dx^2} - 5\frac{dy}{dx} + 6y = 0 \qquad \qquad \textbf{equation 13}$$

$$\frac{dy}{dx} + 2y = 3 \qquad \qquad \textbf{equation 14}$$

$$\frac{d^2y}{dx^2} - 5\frac{dy}{dx} + 6y = \cos 2x \qquad \qquad \textbf{equation 15}$$

Equations **12** and **13** are called **homogeneous** because the right-hand side is zero.

Equations **14** and **15** are called **non-homogeneous** because the right-hand side is a function of x, even though the first is a constant.

Non-homogeneous equations are discussed later in this section.

HOMOGENEOUS LINEAR DIFFERENTIAL EQUATIONS WITH CONSTANT COEFFICIENTS

If you apply the integrating factor method of Section 20.4 to solving the equation $\frac{dy}{dx} + 2y = 0$, you find that the integrating factor is:

$$e^{\int 2\,dx} = e^{2x}$$

When you multiply the equation by this integrating factor, you get

$$e^{2x}\frac{dy}{dx} + 2e^{2x}y = 0$$

which integrates to give $ye^{2x} = A$ where A is a constant.

Therefore the solution of the equation $\frac{dy}{dx} + 2y = 0$ is $y = Ae^{-2x}$ where A is a constant.

The general form of this solution suggests that the equation $\frac{d^2y}{dx^2} - 5\frac{dy}{dx} + 6y = 0$ (equation **15**), might also have a solution of this type, so try $y = Ae^{mx}$, where A and m are non-zero constants, as a solution to see what you can find out about A and m.

As $\frac{dy}{dx} = Ame^{mx}$ and $\frac{d^2y}{dx^2} = Am^2e^{mx}$ you find when you

substitute these into the equation $\frac{d^2y}{dx^2} - 5\frac{dy}{dx} + 6y = 0$ that:

$$Am^2e^{mx} - 5Ame^{mx} + 6Ae^{mx} = 0$$

and, on dividing by e^{mx}, which is never zero, and by A, you find

$$m^2 - 5m + 6 = 0$$

This equation is called the **auxiliary equation**. Its roots are $m = 2$ and $m = 3$ and you can verify that:

$$y = Ae^{2x} + Be^{3x}$$

where A and B are constants is a solution to the equation.

Example 20.5.1

Solve the homogeneous linear equation $\dfrac{d^2y}{dx^2} - 3\dfrac{dy}{dx} - 4y = 0$

The auxiliary equation $m^2 - 3m - 4 = 0$ has roots 4 and -1 so the solution is $y = Ae^{4x} + Be^{-x}$.

Example 20.5.2

Solve the homogeneous linear equation $\dfrac{d^2y}{dx^2} + 4\dfrac{dy}{dx} + 5y = 0$

In this example, the roots of the auxiliary equation $m^2 + 4m + 5 = 0$ are complex, being $-2 + i$ and $-2 - i$.

Assuming a knowledge of complex numbers, the solution is:

$$y = Ae^{(-2 + i)x} + Be^{(-2 - i)x}$$

where the constants A and B may be complex.

You can simplify this equation by writing:

$$y = e^{-2x}(Ae^{ix} + Be^{-ix})$$
$$= e^{-2x}(A\cos x + Ai\sin x + B\cos x - Bi\sin x)$$
$$= e^{-2x}(C\cos x + D\sin x)$$

where $C = A + B$ and $D = i(A - B)$ are also constants.

Example 20.5.3

Solve the equation $\dfrac{d^2y}{dx^2} + 4\dfrac{dy}{dx} + 4y = 0$

You may think that this equation is similar to those of Examples 20.5.1. and 20.5.2, but there is an important difference, because the auxiliary equation $m^2 + 4m + 4 = 0$ has equal roots, in this case -2.

The difficulty arises because when you try to write $y = Ae^{-2x} + Be^{-2x}$ as a solution it simplifies to $y = (A + B)e^{-2x}$ and there is only one constant $A + B$.

You know that there should be two constants, so a solution is missing.

Try $y = ze^{-2x}$ as a solution, where z is a function of x. Then:

$$\frac{dy}{dx} = -2ze^{-2x} + \frac{dz}{dx}e^{-2x}$$

$$\text{and } \frac{d^2y}{dx^2} = \left(\frac{d^2z}{dx^2}e^{-2x} - 2\frac{dz}{dx}e^{-2x}\right) - 2\left(\frac{dz}{dx}e^{-2x} - 2ze^{-2x}\right)$$

$$= \frac{d^2z}{dx^2}e^{-2x} - 4\frac{dz}{dx}e^{-2x} + 4ze^{-2x}$$

After substituting in the equation

$$\frac{d^2y}{dx^2} + 4\frac{dy}{dx} + 4y = 0 \qquad \text{you find} \qquad \frac{d^2y}{dx^2} + 4\frac{dy}{dx} + 4y$$

$$= \frac{d^2z}{dx^2}e^{-2x} - 4\frac{dz}{dx}e^{-2x} + 4ze^{-2x} + 4\left(\frac{dz}{dx}e^{-2x} - 2ze^{-2x}\right) + 4ze^{-2x}$$

$$= e^{-2x}\left(\frac{d^2z}{dx^2}\right) = 0$$

On dividing by e^{-2x} this equation becomes $\frac{d^2z}{dx^2} = 0$ which has a solution $z = Ax + B$, so the solution of the original equation is:

$$y = ze^{-2x} = (Ax + B)e^{-2x}$$

SUMMARY

To solve the homogeneous second order linear differential equation with constant coefficients $\frac{d^2y}{dx^2} + a\frac{dy}{dx} + by = 0$ first construct and solve the auxiliary equation $m^2 + am + b = 0$.

▶ If the auxiliary equation has distinct real roots m_1 and m_2 then the solution is $y = Ae^{m_1x} + Be^{m_2x}$

▶ If the auxiliary equation has complex roots $m_1 + im_2$ and $m_1 - im_2$ then the solution is $y = e^{m_1x}(A\cos m_2x + B\sin m_2x)$

▶ If the auxiliary equation has two roots equal to m, then the solution is $y = (Ax + B)e^{mx}$

NON-HOMOGENEOUS LINEAR DIFFERENTIAL EQUATIONS WITH CONSTANT COEFFICIENTS

To solve equation **14**, $\frac{dy}{dx} + 2y = 3$, first notice by inspection that $y = \frac{3}{2}$ is a solution. Unfortunately there is no arbitrary constant, so the solution is not complete.

However, if you try putting $y = z + \frac{3}{2}$ where z is a function of x, and substituting in the equation $\frac{dy}{dx} + 2y = 3$ you find that:

$$\frac{dz}{dx} + 2z = 0$$

This is a homogeneous equation and you already know its solution, which is $z = Ae^{-2x}$. Thus the whole solution is $y = Ae^{-2x} + \frac{3}{2}$

Notice the form of this solution. It consists of a particular solution $y = \frac{3}{2}$ of the equation $\frac{dy}{dx} + 2y = 3$ added to the general solution of the homogeneous equation $\frac{dy}{dx} + 2y = 0$

Here is a general statement of this result for second order equations.

To solve the equation $\frac{d^2y}{dx^2} + a\frac{dy}{dx} + by = f(x)$, first find a particular solution, called the **particular integral**, of the equation $\frac{d^2y}{dx^2} + a\frac{dy}{dx} + by = f(x)$. Then add the complete solution, called the **complementary function**, of the homogeneous equation $\frac{d^2y}{dx^2} + a\frac{dy}{dx} = by = 0$

This result will be proved later in the section after some examples.

Example 20.5.4

Solve the equation $\frac{d^2y}{dx^2} - 5\frac{dy}{dx} + 6y = 11 - 6x$

First find the particular integral. It is not immediately obvious what this is, so try $y = px + q$ where p and q are constants.

Then $\frac{dy}{dx} = p$ and $\frac{d^2y}{dx^2} = 0$, so by substituting into $\frac{d^2y}{dx^2} - 5\frac{dy}{dx} + 6y$

$$\frac{d^2y}{dx^2} - 5\frac{dy}{dx} + 6y = 0 - 5p + 6(px + q)$$
$$= 6px + (6q - 5p)$$

Identifying $6px + (6q - 5p)$ with $11 - 6x$ gives $p = -1$ and $q = 1$ so that $y = 1 - x$ is the particular integral.

The complementary function is $y = Ae^{2x} + Be^{3x}$ so adding the particular integral to the complementary function to get the general solution gives $y = 1 - x + Ae^{2x} + Be^{3x}$.

Example 20.5.5

Solve the equation $\frac{d^2y}{dx^2} - 5\frac{dy}{dx} + 6y = \cos 2x$

Try $y = p \cos 2x + q \sin 2x$ as the particular integral

Then
$$\frac{dy}{dx} = -2p\sin 2x + 2q \cos 2x$$

$$\frac{d^2y}{dx^2} = -4p \cos 2x - 4q \sin 2x$$

and
$$\frac{d^2y}{dx^2} - 5\frac{dy}{dx} + 6y = -4p \cos 2x - 4q \sin 2x$$
$$- 5(-2p \sin 2x + 2q \cos 2x)$$
$$+ 6(p \cos 2x + q \sin 2x)$$
$$= (2p - 10q) \cos 2x + (2q + 10p) \sin 2x$$

To identify $(2p - 10q)\cos 2x + (2q + 10p)\sin 2x$ with $\cos 2x$ involves solving the simultaneous equations:

$$2p - 10q = 1$$
$$10p + 2q = 0$$

giving $y = \frac{1}{52} \cos 2x - \frac{5}{52} \sin 2x$ as the particular solution.

Adding the complementary function gives the complete solution as

$$y = \frac{1}{52} \cos 2x - \frac{5}{52} \sin 2x + Ae^{2x} + Be^{3x}$$

JUSTIFICATION OF THIS METHOD OF SOLUTION

To justify this method of solution, first suppose that $y = F(x)$ is a particular solution of $\frac{d^2y}{dx^2} = a\frac{dy}{dx} + by = f(x)$ so that:

$$F''(x) + aF'(x) + bF(x) = f(x)$$

and that $y = g(x)$, containing two constants, is a solution of the equation $\frac{d^2y}{dx^2} + a\frac{dy}{dx} + by = 0$ so that:

$$g''(x) + ag'(x) + bg(x) = 0$$

Substituting $y = F(x) + g(x)$ into the expression $\frac{d^2y}{dx^2} + a\frac{dy}{dx} + by$ you find:

$$
\begin{aligned}
\frac{d^2y}{dx^2} + a\frac{dy}{dx} + by &= \left(F''(x) + g''(x)\right) + a\left(F'(x) + g'(x)\right) + b\left(F(x) + g(x)\right) \\
&= \left(F''(x) + aF'(x) + bF(x)\right) + \left(g''(x) + ag'(x) + bg(x)\right) \\
&= f(x) + 0 \\
&= f(x)
\end{aligned}
$$

So $y = F(x) + g(x)$ is a solution. But do you find all the solutions this way?

The answer is yes. For suppose that $y = F_1(x)$ and $y = F_2(x)$ are any two solutions. Consider the function $y = G(x) = F_1(x) - F_2(x)$

You find by substituting this into the expression $\frac{d^2y}{dx^2} + a\frac{dy}{dx} + by$

that it becomes zero. Thus $G(x)$ is a solution of the homogeneous equation. Therefore every possible solution can be obtained from one particular solution by adding the general solution of the homogeneous equation.

Exercise 20.3

Find the general solution of each of the following differential equations. Keep your solutions to questions 1 to 10. They will be useful for questions 11 to 20.

1 $\dfrac{dy}{dx} - 3y = 0$

2 $\dfrac{dy}{dx} + 4y = 0$

3 $\dfrac{d^2y}{dx^2} - 6\dfrac{dy}{dx} + 8y = 0$

4 $\dfrac{d^2y}{dx^2} + 5\dfrac{dy}{dx} - 14y = 0$

5 $\dfrac{d^2y}{dx^2} + 5\dfrac{dy}{dx} = 0$

6 $\dfrac{d^2y}{dx^2} - 4y = 0$

7 $\dfrac{d^2y}{dx^2} + 4y = 0$

8 $\dfrac{d^2y}{dx^2} + 2\dfrac{dy}{dx} + 5y = 0$

9 $\dfrac{d^2y}{dx^2} = 0$

10 $\dfrac{d^2y}{dx^2} + 6\dfrac{dy}{dx} + 9y = 0$

11 $\dfrac{dy}{dx} - 3y = 3$

12 $\dfrac{dy}{dx} + 4y = 4x + 5$

13 $\dfrac{d^2y}{dx^2} - 6\dfrac{dy}{dx} + 8y = 7\cos x + 6\sin x$

14 $\dfrac{d^2y}{dx^2} + 5\dfrac{dy}{dx} - 14y = e^x$ (Try pe^x as the particular integral.)

15 $\dfrac{d^2y}{dx^2} + 5\dfrac{dy}{dx} = 5$

16 $\dfrac{d^2y}{dx^2} - 4y = e^{2x}$ (Try pxe^{2x} as the particular integral.)

17 $\dfrac{d^2y}{dx^2} + 4y = e^{2x}$

18 $\dfrac{d^2y}{dx^2} + 2\dfrac{dy}{dx} + 5y = 10$

19 $\dfrac{d^2y}{dx^2} = 2x + 1$

20 $\dfrac{d^2y}{dx^2} + 6\dfrac{dy}{dx} + 9y = e^{3x}$

20.6 Type V: homogeneous equations

Homogeneous equations are of the form:

$$P + Q\frac{dy}{dx} = 0$$

with P and Q homogeneous functions of the same degree in x and y.

Then $\frac{P}{Q}$ is a function of $\frac{y}{x}$

You can solve homogeneous equations by using the substitution:

$\frac{y}{x} = v$ or $y = vx$.

The two variables x and v then become separable, and you can solve the equation using the method of Section 20.3.

When you have found the solution using the variables x and v, then substitute $\frac{y}{x}$ for v and so reach the final solution.

Example 20.6.1

Solve the differential equation $(x + y) + x\frac{dy}{dx} = 0$

In this example P and Q, that is, $x + y$ and x, are each functions of the first degree throughout in x and y.

Let

$$\frac{y}{x} = v \quad \text{or } y = vx$$

Then

$$\frac{dy}{dx} = v + x\frac{dv}{dx}$$

Substituting in the equation $(x + y) + x\frac{dy}{dx} = 0$

$$x + vx + x\left(v + x\frac{dv}{dx}\right) = 0$$

or
$$1 + 2v + x\frac{dv}{dx} = 0$$

Separating the variables

$$\frac{dv}{1 + 2v} = -\frac{dx}{x}$$

Integrating

$$\tfrac{1}{2}\ln|1 + 2v| = -\ln|x| + \tfrac{1}{2}\ln c$$

where, for convenience, the constant has been written as $\tfrac{1}{2}\ln c$

This leads to

$$\ln|1 + 2v| = -2\ln|x| + \ln c$$

or
$$x^2(1 + 2v) = c$$

Substituting for v gives

$$x^2\left(1 + 2\frac{y}{x}\right) = c$$

Therefore the solution is $x^2 + 2xy = c$

Example 20.6.2

Solve the equation $(x^2 - y^2)\dfrac{dy}{dx} = 2xy$

Put $y = vx$, so $\dfrac{dy}{dx} = v + x\dfrac{dv}{dx}$

Substituting,

$$(x^2 - v^2x^2)\left(v + x\frac{dv}{dx}\right) = 2vx^2$$

Dividing by x^2 gives

$$x(1 - v^2)\frac{dv}{dx} = v + v^3$$

Separating the variables,

$$\frac{1-v^2}{v+v^3} dv = \frac{dx}{x}$$

Using partial fractions and integrating

$$\int \frac{1-v^2}{v+v^3} dv = \int \frac{dx}{x}$$

$$\int \left(\frac{1}{v} - \frac{2v}{1+v^2} \right) dv = \int \frac{dx}{x}$$

$$\ln|v| - \ln|1+v^2| = \ln|x| + \ln c$$

$$\frac{v}{1+v^2} = cx$$

Replacing v by $\frac{y}{x}$

$$\frac{\dfrac{y}{x}}{1+\left(\dfrac{y}{x}\right)^2} = cx$$

leads to

$$\frac{xy}{x^2+y^2} = cx$$

The solution is $y = c(x^2 + y^2)$

Test yourself

Solve the following differential equations.

1 $x + y + x\dfrac{dy}{dx} = 0$

2 $x + y - x\dfrac{dy}{dx} = 0$

3 $x + y + (y - x)\dfrac{dy}{dx} = 0$

4 $x - 2y + y\dfrac{dy}{dx} = 0$

5 $x^2 + y^2 = 2xy\dfrac{dy}{dx}$

6 $y^2 - x^2\dfrac{dy}{dx} = 0$

7 $y^2 - 2xy = (x^2 - 2xy)\dfrac{dy}{dx}$

8 $x^2\dfrac{dy}{dx} + y^2 + xy\dfrac{dy}{dx} = 0$

9 $y^2 + (x^2 - xy)\dfrac{dy}{dx} = 0$

10 $y^2 + (x^2 + xy)\dfrac{dy}{dx} = 0$

11 $x^2\dfrac{dy}{dx} + xy + 1 = 0$

Applications of differential equations

In this chapter you will learn:

▶ *how to set up differential equations to describe physical phenomena*

▶ *see how differential equations are applied to the cooling of a liquid, the motion of a rocket and to the curve formed by a hanging chain.*

21.1 Introduction

Nugget – But what's the point of differential equations?

A feature of the world around us is that it is constantly changing. Because rates of change can be represented mathematically by derivatives, you will not be surprised to learn that many of the models of growth and change in fields such as biology, economics and physics employ the language of derivatives to describe the changes being investigated. Examples include the growth of an organism, the fluctuations of the stock market, the spread of diseases, the rate of change of an urban population, weather prediction, how food moves around the intestines, and much more.

When you tackle a problem which includes a differential equation as part of its solution, you must, somewhere in the process, have first set up the differential equation. You may be able to do this immediately from the given information. For example if you know that the rate of change of some quantity q, at a given instant, is directly proportional to the quantity existing at that instant then you can write immediately that

$$\frac{dq}{dt} = kq$$

where k is a constant. You can then solve this equation – it is variables separable – to find that

$$\int \frac{dq}{q} = \int k \, dt$$

so $\ln q = kt + \ln c$ or $\ln\left(\frac{q}{c}\right) = kt$ giving $q = Ae^{kt}$

where A is a constant. If you know the initial conditions, then you can find the value of A and the problem is solved.

Another type of differential equation is one in which you need to work from first principles and think about a small element of some kind.

Examples of these types of process are given in the next two sections.

21.2 Problems involving rates

The two examples in this section are typical of situations where you are given information which you can rewrite in the form of differential equations.

Example 21.2.1

The acceleration of a particle is proportional to its displacement from a fixed point in its path, and of opposite sign. Find equations for its velocity and displacement as a function of time.

Let x metres be its displacement at time t seconds. Then, from the information given:

$$\frac{d^2x}{dt^2} = -kx$$

where k is a positive constant. Then:

$$\frac{d^2x}{dt^2} + kx = 0$$

which is a linear differential equation with constant coefficients. (See Section 20.5.) Its solution is:

$$x = A\cos\sqrt{k}t + B\sin\sqrt{k}t$$

Suppose that the initial conditions are that when $t = 0$, $x = 0$, and that when its velocity $v = 0$, $x = a$.

Then putting $t = 0$, $x = 0$ into the equation $x = A\cos\sqrt{k}t + B\sin\sqrt{k}t$ you find $0 = A$, giving the equation:

$$x = B\sin\sqrt{k}t$$

Its velocity v is given by

$$v = \frac{dx}{dt} = \sqrt{k}B\cos\sqrt{k}t$$

Since for any angle θ, $\sin^2\theta + \cos^2\theta = 1$,

$$\frac{x^2}{B^2} + \frac{v^2}{kB^2} = 1$$

so

$$kx^2 + v^2 = kB^2$$

Using the fact that when $v = 0$, $x = a$

$$ka^2 = kB^2$$

so

$$B = a$$

Therefore the required equations are $x = a\sin\sqrt{k}t$ and $v = \sqrt{k}a\cos\sqrt{k}t$

This kind of motion is called **simple harmonic motion** and has application in all kinds of wave motion.

Example 21.2.2

Newton's Law of Cooling states that under certain conditions the rate of loss of temperature of a body is proportional to the excess of temperature of the body over the ambient temperature of the surroundings. In an experiment, a body which starts at temperature 90°C falls to 80°C in 10 minutes and to 75°C in 20 minutes. Find the ambient temperature.

Let the temperature of the body at time t minutes be T°C, and let the ambient temperature be θ°C, where θ is a constant.

Then Newton's Law of Cooling states that:

$$\frac{dT}{dt} = -k(T - \theta)$$

or, since θ is a constant

$$\frac{dT}{dt} + kT = k\theta$$

The solution of this differential equation is:

$$T = \theta + Ae^{-kt}$$

From the initial conditions

$$90 = \theta + A$$
$$80 = \theta + Ae^{-10k}$$
$$75 = \theta + Ae^{-20k}$$

Therefore
$$\frac{80 - \theta}{90 - \theta} = e^{-10k}$$

and
$$\frac{75 - \theta}{90 - \theta} = e^{-20k}$$

so
$$\frac{75 - \theta}{90 - \theta} = \left(\frac{80 - \theta}{90 - \theta}\right)^2$$

Therefore $\qquad (75 - \theta)(90 - \theta) = (80 - \theta)^2$

which gives $\qquad\qquad \theta = 70$

Therefore the ambient temperature of the surroundings is 70°C.

21.3 Problems involving elements

In the next two examples progress is made by considering a small element. Both examples require some knowledge of mechanics, and are fairly demanding.

Example 21.3.1

Investigate the motion of a rocket travelling vertically.

When a rocket travels upwards, it propels itself by burning fuel which is gradually used up, making the rocket lighter. The principle of mechanics that applies is Newton's Second Law. This states that the rate of change of momentum is proportional to the external force.

Suppose that a rocket of mass M carries an additional mass m of fuel. Suppose also that the fuel burns at a constant rate μ and is ejected from the back of the rocket at a constant speed V relative to the rocket, until the fuel is used up. Suppose also that the rocket starts from rest.

Let v be the speed of the rocket at time t, and consider what happens in the next element of time δt. This situation is shown in Figure 21.1.

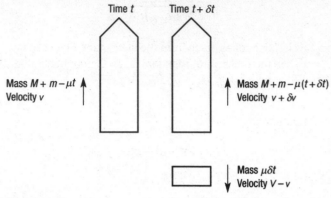

Mass $M + m - \mu t$
Velocity v

Mass $M + m - \mu(t + \delta t)$
Velocity $v + \delta v$

Mass $\mu \delta t$
Velocity $V - v$

Figure 21.1 External force acting on the system is $(M + m - \mu t)g$

The change in momentum between time $t + \delta t$ and t is:

$$[\{M + m - \mu(t + \delta t)\}(v + \delta v) - \mu \delta t(V - v)] - (M + m - \mu t)v$$
$$= (M + m - \mu t)\delta v - \mu \delta t V$$

The rate of change of momentum upwards is therefore given by:

$$\lim_{\delta t \to 0}\left(\frac{\text{change of momentum in time } \delta t}{\delta t}\right)$$

$$= \lim_{\delta t \to 0}\left\{\frac{(M + m - \mu t)\delta v - \mu \, \delta t \, V}{\delta t}\right\}$$

$$= \lim_{\delta t \to 0}\left\{(M + m - \mu t)\frac{dv}{dt} - \mu V\right\}$$

$$= (M + m - \mu t)\frac{dv}{dt} - \mu V$$

The external force is $(M + m - \mu t)g$ downwards.

Therefore, using Newton's Second Law

$$(M + m - \mu t)\frac{dv}{dt} - \mu V = -\left((M + m - \mu t)\right)g$$

Rearranging this equation gives

$$\frac{dv}{dt} = \frac{\mu V}{(M + m - \mu t)} - g$$

This acceleration takes place from the start of the fuel burning at $t = 0$ when the rocket is at rest, that is, $v = 0$, until the fuel is exhausted at time $t = \frac{m}{\mu}$

Integrating the equation $\dfrac{dv}{dt} = \dfrac{\mu V}{M + m - \mu t} - g$

$$v = -V \ln(M + m - \mu t) - gt + C$$

Using the condition that $v = 0$ when $t = 0$

$$C = V \ln(M + m)$$

Therefore
$$v = V \ln\left(\frac{M + m}{M + m - \mu t}\right) - gt$$

When the fuel is exhausted, that is, when $t = \frac{m}{\mu}$, the final velocity v_F is given by:

$$v_F = V \ln\left(\frac{M + m}{M}\right) - \frac{gm}{\mu}$$

It is interesting to take some typical values of the constants in this equation. Assume that the mass m of fuel is about $4M$ and that $v = 3000$ ms^{-1} so that the final velocity v_F is given by:

$$v_F \approx 4800 - \frac{gm}{\mu} \text{ms}^{-1}$$

It should not be a surprise that, the faster the rate of burning (that is, the greater the value of μ), the higher the final velocity.

Example 21.3.2

What is the shape of the curve taken by a heavy, flexible, uniform rope hanging stretched between two points?

Let $P(x, y)$ be a point on the curve, and consider a neighbouring point Q with coordinates $(x + \delta x, y + \delta y)$. Let the angles which the rope makes with the x-axis at P and Q be ψ and $\psi + \delta\psi$. Let the length PQ of the rope be δs, and let the tension in the rope at P be T, and the tension at Q be $T + \delta T$. Let the rope have a weight of w per unit length.

This small section of rope is shown in Figure 21.2.

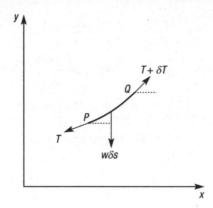

Figure 21.2

Consider the forces acting on the piece of rope.

Horizontally, the only forces acting are the horizontal components of T and $T + \delta T$, that is, $T\cos\psi$ and $(T + \delta T)\cos(\psi + \delta\psi)$.

As the rope is in equilibrium, these two forces are equal, so:

$$T\cos\psi = (T + \delta T)\cos(\psi + \delta\psi)$$

Therefore $T\cos\psi$ does not change along the length of the rope, so $T\cos\psi$ must be constant. Call this constant wc, where c is constant.

Therefore $\qquad\qquad T\cos\psi = wc$

Vertically, the forces acting are the vertical components of the tension, which must balance the weight of the piece of rope.

Therefore $\qquad (T + \delta T)\sin(\Psi + \delta\Psi) - T\sin\Psi = w\delta s$

Expanding the left-hand side and dividing by δs

$$\left(\frac{T}{\delta x} + \frac{\delta T}{\delta s}\right)\sin\psi\cos\delta\psi + (T + \delta T)\cos\psi\frac{\sin\delta\psi}{\delta s} - \frac{T\sin\psi}{\delta s} = w$$

As ψ is in radians, and $\delta\psi$ is small you can use Maclaurin's series (Section 19.4), to show that $\cos\delta\psi \approx 1 - \frac{1}{2}(\delta\psi)^2$.

Remember also that $\dfrac{\sin\delta\psi}{\delta s} = \dfrac{\sin\delta\psi}{\delta\psi} \times \dfrac{\delta\psi}{\delta s}$, so

$$\left(\frac{T}{\delta s} + \frac{\delta T}{\delta s}\right)\sin\psi\left(1 - \tfrac{1}{2}(\delta\psi)^2\right) + (T + \delta T)\cos\psi\frac{\sin\delta\psi}{\delta\psi}.\frac{\delta\psi}{\delta s} - \frac{T\sin\psi}{\delta s} = w$$

Multiplying out the brackets and taking the limit as $\delta s \to 0$

$$\frac{dT}{ds}\sin\psi + T\cos\psi\frac{d\psi}{ds} = w$$

which is the same as

$$\frac{d}{ds}(T\sin\psi) = w$$

which, on integration gives $T\sin\psi = ws + A$ where A is constant

Measure s so that $s = 0$ when $\psi = 0$, so $A = 0$

Therefore $T\cos\psi = wc$ and $T\sin\psi = ws$

Dividing these equations gives

$$s = c\tan\psi$$

or, since $\tan\psi = \dfrac{dy}{dx}$

$$s = c\frac{dy}{dx}$$

Let $\frac{dy}{dx} = p$, and differentiate this equation with respect to x. Then

$\frac{ds}{dx} = c\frac{dp}{dx}$, or using the result of Section 18.1:

$$c\frac{dp}{dx} = \sqrt{1 + \left(\frac{dy}{dx}\right)^2} = \sqrt{1 + p^2}$$

Therefore $$c\frac{dp}{dx} = \sqrt{1 + p^2}$$

This equation is a variables separable equation (see Section 20.3) which you can solve to get:

$$x = c\sinh^{-1}p + B$$

where B is constant.

If you measure $x = 0$ when $p = \frac{dy}{dx} = 0$, then $B = 0$

Therefore $\qquad\qquad\qquad\qquad\qquad p = \sinh\frac{x}{c}$

Since $p = \frac{dy}{dx}$, $\frac{dy}{dx} = \sinh\frac{x}{c}$ which you can integrate to get $y = c\cosh\frac{x}{c} + C$ where C is constant. If you choose the origin so that when $x = 0$, $y = c$, you find that $C = 0$.

Therefore the equation of the hanging rope is $y = c\cosh\frac{x}{c}$

This curve is called a **catenary**, which comes from the Latin word *catena*, a chain. The graph of the catenary $y = \cosh x$ is shown in Figure 21.3.

The catenary is similar in shape to the parabola, but it is somewhat flatter at the bottom.

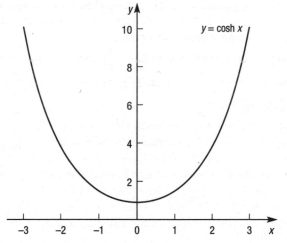

Figure 21.3

Test yourself

1 The velocity in metres per second of a stopping train as it moves between stations is given by

$$v = \frac{t(60 - t)}{45}$$

Calculate the distance between the stations.

2 Radium is radioactive. When radioactive substances decay, the number of atoms that decay in a fixed time period is proportional to the number of atoms at the start of that period. Radium is known to decay so that, after 1600 years the amount of radium present is halved. Suppose that there are initially 10 grams of radium. Find how much radium is left after 100 years.

3 A kilogram of toxic material has been spilt into a tank containing 1000 litres of water and is thoroughly mixed. One way of getting rid of the toxic material is to allow the mixture to run out from a tap at the bottom of the tank, and to refill the tank at the same rate with water. The mixture runs out at a rate of 20 litres per minute. Find when only 0.1 kilograms of toxic material remain.

4 The following proposal for describing population growth was put forward by the Belgian demographer P. F. Verhulst in the 1830s. The rate of growth of a population with p inhabitants is proportional to $\frac{P}{A}\left(1 - \frac{P}{A}\right)$ where A is a constant. Show that if the initial population p_0 has the property that $0 < p_0 < A$, then $\lim_{t \to \infty} p = A$

5 Suppose that a load is suspended from a hanging chain in such a way that the load is proportional to the horizontal projection of the chain. A suspension bridge is an example of such a situation, the load being so much heavier than the weight of the cable, that the weight of the cable can be neglected in comparison with the road. Use the notation of Example 21.3.2 to show that the shape of the cable is the parabola $y = \frac{x^2}{2c}$

Answers

Chapter 1: Test yourself

1 $-1, 1, 1, 17, 2a^2 - 4a + 1, 2(x + \delta x)^2 - 4(x + \delta x) + 1$

2 $7, 0, -5, a(a + 6), \dfrac{(1 - a)(1 + 5a)}{a^2}, 0$

3 $0, 1, \frac{1}{2}, \frac{1}{2}\sqrt{3}, -1$

4 $9, 9.61, 9.0601, 9.006001, 6.001$

5 $1, 2, 8, 1.414\ldots$

6 $7, 0, -11, -x^3 - 5x^2 + 3x + 7$

7 $3(t + \delta t)^2 + 5(t + \delta t) - 1$

8 $2x.\delta x + 2\delta x + (\delta x)^2$

9 **a** $x^3 + 3x^2.\delta x + 3x(\delta x)^2 + (\delta x)^3$ **b** $3x^2.\delta x + 3x(\delta x)^2 + (\delta x)^3$
   c $3x^2 + 3x.\delta x + (\delta x)^2$

10 **a** $2x^2 + 4hx + 2h^2$ **b** $4hx + 2h^2$ **c** $4x + 2\delta x$

Chapter 2: Test yourself

1 **a** 0 **b** Values less than 1 **c** $1, 1.25, 2, 5, 10, -2, -1,$
   $-\frac{1}{2}, -\frac{1}{3}$

   d Infinity **e** The graph is a hyperbola similar to that of
   Figure 2.1, but the y-axis is at $x = 1$.

2 **a** $3.1, 3.01, 3.001, 3.000001$ **b** 3

3 **a** 5 **b** Infinity

4 **a** $11, 5, 3, 2.5, 2.1, 2.01$ **b** 2

5 Infinity

6 2

7 $3x^2$

8 $\frac{1}{2}$

9 $\frac{4}{3}$

Chapter 3: Test yourself

1 1.5, 1.2

2 a 2.5   b –0.8   c $-\frac{b}{a}$

3 $y = 1.2x + 4$

4 6

5 $\delta y = 3a^2 \times \delta x + 3a(\delta x)^2 + (\delta x)^3, \dfrac{\delta y}{\delta x} = 3a^2 + 3a \times \delta x + (\delta x)^2, 12$

6 $\delta y = \dfrac{-\delta x}{a^2 + a.\delta x}, \dfrac{\delta y}{\delta x} = \dfrac{-1}{a^2 + a.\delta x}, -1, 135°$

7 a 2   b 2

8 a 12   b 8

Chapter 4: Test yourself

1 a 12 ms^{-1} b 6 ms^{-1} c 18 ms^{-1}

2 a 2   b 2

3 a 0   b –1   c –1.9   d $\delta x - 2$. The rate of change is –2. The velocity at $t = 0$ is –2 ms^{-1}

4 $\delta s = 9.8T \times \delta t + 4.9(\delta t)^2, \dfrac{\delta s}{\delta t} = 9.8T + 4.9\delta t$

   a 20.58   b 20.09   c 19.649   d 19.6049. The velocity is 19.60 ms^{-1}

Chapter 5: Test yourself

1 $7x^6, 5, \frac{1}{3}, 0.06, \frac{5}{4}x^4, 60x^3, 4x^5, 4.5x^2, 32x$

2 $4bx^3, \dfrac{6ax^5}{b}, apx^{p-1}, 2ax^{2a-1}, 2(2b+1)x^{2b}, 8\pi x$

3 $6, 0.54, -3, p$

4 In each part of this question a constant could be added to the answer. $\frac{1}{2}x^2, \frac{3}{2}x^2, \frac{1}{3}x^3, \frac{1}{12}x^3, \frac{1}{6}x^6, \dfrac{x^{n+1}}{n+1}, \dfrac{x^{2a+1}}{2a+1}, \dfrac{x^4}{6}, \dfrac{4ax^3}{3}$

5 a

6 $20t$

7 $2\pi r$

8 $4\pi r^2$

9 $\dfrac{5}{2\sqrt{x}}$ $\dfrac{-5}{x^2}$ $\dfrac{-5}{2x^{\frac{3}{2}}}$ $\dfrac{2}{3\sqrt[3]{x}}$ $\dfrac{3}{4}\sqrt[4]{\dfrac{2}{x}}$

10 $\dfrac{0.4}{x^{0.6}}$ $\dfrac{1.6}{x^{0.8}}$ $-\dfrac{1.6}{x^{1.2}}$ $-\dfrac{24}{x^5}$ $-\dfrac{p}{x^{p+1}}$

11 $19.2x^{2.2}$ $\dfrac{-3}{x^{2.5}}$ $\dfrac{20.3}{x^{0.3}}$ $\dfrac{-18}{5x^{\frac{8}{5}}}$

12 $\dfrac{-40}{v^3}$

13 $1.5, 0$

14 24

15 $-0.02, -0.5, -2, -8$

16 $\dfrac{-2}{x^3}$

17 $x = 1$, i.e. $(1, 1)$

18 $x = \dfrac{1}{\sqrt{3}}$ or $x = -\dfrac{1}{\sqrt{3}}$ i.e. $\left(\dfrac{1}{\sqrt{3}}, \dfrac{1}{3\sqrt{3}}\right)$ and $\left(\dfrac{-1}{\sqrt{3}}, \dfrac{-1}{3\sqrt{3}}\right)$

19 $x = \frac{1}{16}$ i.e. $\left(\frac{1}{16}, \frac{1}{4}\right)$

20 $x = \frac{1}{2}$ i.e. $\left(\frac{1}{2}, \frac{1}{8}\right)$

Chapter 6: Exercise 6.1

1 a $12x + 5$ **b** $9x^2 + 1$ **c** $16x^3 + 6x - 1$ **d** $x + \frac{1}{7}$

 e $-\dfrac{5}{x^2} + 4$ **f** $-\dfrac{4}{x^2} + \dfrac{4}{x^3}$ **g** $5 - 2x + 9x^2$ **h** $\dfrac{4}{\sqrt{x}}$

2 a $u + at$ **b** $5 + 32t$ **c** $6t - 4$

3 $3ax^2 + 2bx + c$

4 $2x - \dfrac{2}{x^3}$

5 $\dfrac{1}{2\sqrt{x}}\left(1 - \dfrac{1}{x}\right)$

6 $3(1 + x)^2$

7 $2nx^{2n-1} - 2nx + 5$

8 $\dfrac{1}{2\sqrt{x}} + \dfrac{1}{3\sqrt[3]{x^2}} - \dfrac{2}{x^2}$

9 $3, x = \frac{3}{4}$

10 2 and -2

11 $2, -1, 2$

12 $(1, 2)$ and $(-1, -2)$

Exercise 6.2

1 $12x + 5$

2 $\frac{3}{2}x^2 + 2x + \frac{1}{2}$

3 $9x^2 + 2x - 10$

4 $8x^3 + 10x$

5 $12x^3 + 33x^2 - 8x$

6 $3x^2$

7 $3x^2$

8 $4x^3 + 12x^2 + 6x - 8$

9 $4x^3$

10 $4x^3 - 2x + 2$

11 $3x^2$

12 $24x^3 + 6x^2 - 22x - 3$

13 $18x^2 + 2x - 5$

14 $(2ax + b)(px + q) + p(ax^2 + bx + c)$

15 $3\sqrt{x}\left(\sqrt{x} + 2\right)\left(\sqrt{x} - 1\right) + x\left(2\sqrt{x} + 1\right)$

Exercise 6.3

1 $\dfrac{-6}{(2x - 1)^2}$

2 $\dfrac{6x}{(1 - 3x^2)^2}$

3 $\dfrac{2}{(x + 2)^2}$

4 $\dfrac{1}{(x + 2)^2}$

5 $\dfrac{11}{(2x + 3)^2}$

6 $\dfrac{-2b}{(x - b)^2}$

7 $\dfrac{2b}{(x + b)^2}$

8 $\dfrac{x^2 - 8x}{(x - 4)^2}$

9 $\dfrac{-8x}{(x^2 - 4)^2}$

10 $\dfrac{1 - x}{2\sqrt{x}(x + 1)^2}$

11 $\dfrac{x - 1}{2x^{\frac{3}{2}}}$

12 $\dfrac{-1}{\sqrt{x}\left(\sqrt{x} - 1\right)^2}$

13 $\dfrac{6x^2}{(x^3 + 1)^2}$

14 $\dfrac{-2x^2 + 4x}{(x^2 - x + 1)^2}$

15 $\dfrac{5x^2 - 10x}{(3x^2 + x - 1)^2}$

16 $\dfrac{x^2 - 1}{x^2}$

17 $\dfrac{4x^3(2a^2 - x^2)}{(a^2 - x^2)^2}$

18 $\dfrac{-5}{(2 - 3x)^2}$

19 $\dfrac{x^2 - 4x + 2}{x^2 - 4x + 4}$

20 $\dfrac{-\left(x + 3x^{\frac{1}{2}}\right)}{x^3}$

Exercise 6.4

1 $4(2x + 5)$

2 $-20(1 - 5x)^3$

3 $(3x + 7)^{-\frac{2}{3}}$

4 $\dfrac{2}{(1 - 2x)^2}$

5 $-4(1 - 2x)$

6 $\dfrac{-1}{\sqrt{1 - 2x}}$

7 $10x(x^2 - 4)^4$

8 $-3x\sqrt{1 - x^2}$

9 $\dfrac{3x}{\sqrt{3x^2 - 7}}$

10 $\dfrac{4x}{(1 - 2x^2)^2}$

11 $\dfrac{-2x}{\sqrt{1 - 2x^2}}$

12 $\dfrac{1 - 2x^2}{\sqrt{1 - x^2}}$

13 $\dfrac{1}{(4 - x)^2}$

14 $\dfrac{1}{2(4 - x)^{\frac{3}{2}}}$

15 $\dfrac{2}{(4 - x)^3}$

16 $\dfrac{-2x}{(x^2 - 1)^2}$

17 $\dfrac{-x}{(x^2 - 1)^{\frac{3}{2}}}$

18 $\dfrac{1}{(1 + x^2)^{\frac{3}{2}}}$

19 $\dfrac{1}{(1 - x^2)^{\frac{3}{2}}}$

20 $\dfrac{1}{2\sqrt{\{x(1 - x)^3\}}}$

21 $\dfrac{-1}{(1 + x)^{\frac{3}{2}}(1 - x)^{\frac{1}{2}}}$

22 $\dfrac{1 - x - x^2}{(1 + x)^{\frac{3}{2}}(1 - x)^{\frac{1}{2}}}$

23 $\dfrac{2x}{3(x^2 + 1)^{\frac{2}{3}}}$

24 $\dfrac{x}{\sqrt{a^2 + x^2}}$

25 $\dfrac{-x}{(a^2 + x^2)^{\frac{3}{2}}}$

26 $\dfrac{2x - 1}{2\sqrt{1 - x + x^2}}$

27 $-4nx(1 - 2x^2)^{n-1}$

28 $\dfrac{x(2a^2 - x^2)}{(a^2 - x^2)^{\frac{3}{2}}}$

29 $2\left(x + \dfrac{1}{x}\right)\left(1 - \dfrac{1}{x^2}\right)$

30 $\dfrac{-3x^2}{2(1 + x^3)^{\frac{3}{2}}}$

31 $\dfrac{x - 1}{x^2\sqrt{1 + 2x}}$

32 $\dfrac{1}{(1 + x^2)^{\frac{3}{2}}}$

33 $\dfrac{1}{x^2\sqrt{1 + x^2}}$

34 $\dfrac{-4x + 3}{2(2x^2 - 3x + 4)^{\frac{3}{2}}}$

35 $\dfrac{4x - 5x^2}{2\sqrt{1 - x}}$

36 $\dfrac{1}{(1 - x)\sqrt{1 - x^2}}$

37 $\dfrac{3(x + 1)}{\sqrt{2x + 3}}$

Exercise 6.5

1   $\dfrac{-6x - 7y}{7x + 18y}$ 2 $-\dfrac{2x(x^2 + y^2) - x}{2y(x^2 + y^2) + y}$ 3 $\dfrac{y - x^2}{y^2 - x}$

4   $-\dfrac{x^{n-1}}{y^{n-1}}$ 5 $\dfrac{dy}{dx} = \dfrac{-2x + 3}{2y + 4} = \dfrac{1}{6}$ at $(1,\ 1)$.

Exercise 6.6

1   $x(3x - 2)$, $2(3x - 1)$, 6

2   $2bx^{2b-1}$, $2b(2b - 1)x^{2b-2}$, $2b(2b - 1)(2b - 2)x^{2b-3}$

3   $20x^3 - 9x^2 + 4x - 1$, $60x^2 - 18x + 4$, $120x - 18$

4   $50x^4 - 12x^2 + 5$, $200x^3 - 24x$, $600x^2 - 24$

5   $-\dfrac{1}{x^2}$ $\dfrac{2}{x^3}$ $-\dfrac{6}{x^4}$

6   $\dfrac{1}{2x^{\frac{1}{2}}}$ $\dfrac{-1}{4x^{\frac{3}{2}}}$ $\dfrac{3}{8x^{\frac{5}{2}}}$

7   $\dfrac{1}{\sqrt{2x + 1}}$ $\dfrac{-1}{\sqrt{(2x + 1)^3}}$ $\dfrac{3}{\sqrt{(2x + 1)^5}}$

8   $-\dfrac{2}{x^3}$ $\dfrac{6}{x^4}$ $-\dfrac{24}{x^5}$

9   $\dfrac{n!}{2a}\left\{\dfrac{1}{(a - x)^{n+1}} + \dfrac{(-1)^n}{(a + x)^{n+1}}\right\}$

10   $-5, \frac{5}{12}$, the lowest point of the curve

11   $-7, 2, 0$ and $\frac{10}{3}$

12   $x = 3$, $x = 2$, 2.5, -0.25 (the turning value)

Chapter 7: Test yourself

1   $\dfrac{dy}{dx} = 2x - 2$; $-4, -2, 0, 2, 4$; $x = 1$; local minimum; positive.

2   $\dfrac{dy}{dx} = 3 - 2x$, $3, 1, -1, -3$; 1.5; negative; local maximum.

3 a $x = \frac{1}{4}$, minimum, at $\left(\frac{1}{4}, -\frac{1}{4}\right)$

  b $x = \frac{1}{3}$, maximum, at $\left(\frac{1}{3}, \frac{1}{6}\right)$

  c $x = -2$, minimum, at $(-2, -2)$

  d $x = -\frac{1}{4}$, minimum, at $\left(-\frac{1}{4}, -\frac{9}{8}\right)$

4 a Minimum -16 at $x = 2$, maximum 16 at $x = -2$.

  b Maximum 5 at $x = 1$, minimum 4 at $x = 2$.

  c Maximum 12 at $x = 0$, minimum -20 at $x = 4$.

  d Maximum 41 at $x = -2$, minimum $9\frac{3}{4}$ at $x = \frac{1}{2}$.

5 Maximum 4 at $x = 0$, minimum 0 at $x = 2$.

6 Maximum -4 at $x = -\frac{1}{2}$, minimum 4 at $x = \frac{1}{2}$.

7 $5, 5$

8 $\dfrac{u^2 \sin^2 \theta}{2g}, \dfrac{u^2 \sin 2\theta}{2g}$ or $\dfrac{u^2 \sin \theta \cos \theta}{g}$

9 Height : diameter $= 1 : 1$

10 The square base has side 2.52 m; the height is 1.26 m.

11 $s = 3 + 4.8t - 1.6t^2$; 6.6

12 4.5

13 $9 - \sqrt{21} = 4.42$ cm approximately.

14 1.5 m deep, 0.5 m broad.

15 $x = -3, x = 2, x = -\frac{1}{2}$

16 Maximum 0.385, minimum -0.385; gradient $= -1$

17 $(0, 5)$

18 Height 734.69, time 12.24

19 Centre of the beam.

Chapter 8: Exercise 8.1

1 $3\cos x$

2 $3\cos 3x$

3 $-\frac{1}{2}\sin \frac{1}{2}x$

4 $\frac{1}{3}\sec^2 \frac{1}{3}x$

5 $0.6 \sec 0.6x \tan 0.6x$

6 $-\frac{1}{6}\operatorname{cosec}\frac{1}{6}x \cot \frac{1}{6}x$

7 $2(\cos 2x - 2\sin 2x)$

8 $3(\cos 3x + \sin 3x)$

9 $\sec x(\sec x + \tan x)$

10 $4\cos 4x - 5\sin 5x$

11 $-\frac{1}{2}\sin\frac{1}{2}\theta + \frac{1}{4}\cos\frac{1}{4}\theta$

12 $2\cos\left(2x + \frac{1}{2}\pi\right)$

13 $\sin(3\pi - x)$

14 $\frac{1}{2}\operatorname{cosec}\left(a - \frac{1}{2}x\right)\cot\left(a - \frac{1}{2}x\right)$

15 $3\sin^2 x \cos x$

16 $3x^2 \cos x^3$

17 $-6\cos^2(2x)\sin(2x)$

18 $2x\sec x^2 \tan x^2$

19 $\dfrac{-\sec^2\left(\sqrt{1-x}\right)}{2\sqrt{1-x}}$

20 $\sec^2\frac{x}{2}$

21 $a\sin x$

22 $n(a\cos nx - b\sin nx)$

23 $-2\sin\left(2x + \frac{1}{2}\pi\right)$

24 $2\sec^2 2x - 2\tan x \sec^2 x$

25 $2x + \frac{3}{2}\cos\frac{1}{2}x$

26 $x\cos x + \sin x$

27 $\frac{a}{x^2}\sin\frac{a}{x}$

28 $\dfrac{\sin x - x\cos x}{\sin^2 x}$

29 $x\sec^2 x + \tan x$

30 $2\cos 2x + 8x\cos(2x)^2$

31 $\dfrac{x\sec^2 x - \tan x}{x^2}$

32 $\dfrac{\tan x - x\sec^2 x}{\tan^2 x}$

33 $-6x\cos^2(x^2)\sin(x^2)$

34 $2x\tan x + x^2\sec^2 x$

35 $-5\operatorname{cosec}^2(5x + 1)$

36 $-6\cot 3x \operatorname{cosec}^2 3x$

37 $\dfrac{-\sin x}{2\sqrt{\cos x}}$

38 $2(\cos^2 2x - \sin^2 2x)$

39 0

40 $4\sin x \cos x$

41 $\dfrac{\sin x}{(1 + \cos x)^2}$

42 $\dfrac{2\sin x}{(1 + \cos x)^2}$

43 $\dfrac{\sin x - 2x\cos x}{2\sqrt{x}\sin^2 x}$

44 $\dfrac{\sec^2 x}{(1 - \tan x)^2}$

45 $\dfrac{2x(\cos 2x + x\sin 2x)}{\cos^2 2x}$

46 $\sin x + \cos x$

47 $\dfrac{2\sin x + x\cos x}{2\sqrt{\sin x}}$

48 $\dfrac{\sin x \cos x(2 + \sin x)}{(1 + \sin x)^2}$

49 $2x\cos 2x - 2x^2\sin 2x$

50 $\sec^2 x \operatorname{cosec} x\,[2\tan x - \cot x]$

Exercise 8.2

1 Maximum, $x = \frac{1}{6}\pi$; minimum, $x = \frac{5}{6}\pi$

2 Maximum, $x = \frac{1}{4}\pi$ and $\frac{3}{4}\pi$; minimum, $x = \frac{1}{2}\pi$

3 Maximum, $x = \frac{1}{3}\pi$

4 Maximum, $x = \frac{1}{4}\pi$

5 Maximum, $x = \tan^{-1}2$

6 Maximum, $x = \frac{1}{4}\pi$

7 Maximum, $x = \frac{2}{3}\pi$; minimum, $x = \frac{4}{3}\pi$

8 Maximum, $x = \pm \sin^{-1}\sqrt{2/3}$ or $\pm \cos^{-1}(\frac{1}{\sqrt{3}})$ or $x = 0$;

   minimum, $x = 0$

9 $x = \tan^{-1}\left(\frac{3}{2}\right)$

10 Minimum -1, $x = \frac{1}{4}\pi$

11 $\dfrac{4}{\sqrt{1 - 16x^2}}$

12 $\dfrac{1}{\sqrt{4 - x^2}}$

13 $\dfrac{-b}{\sqrt{a^2 - x^2}}$

14 $\dfrac{-1}{\sqrt{9 - x^2}}$

15 $\dfrac{3}{9 + x^2}$

16 $\dfrac{-1}{1 + (a - x)^2}$

17 $\dfrac{-4x}{\sqrt{1 - 4x^4}}$

18 $\dfrac{1}{2\sqrt{x - x^2}}$

19 $\sin^{-1}x + \dfrac{x}{\sqrt{1 - x^2}}$

20 $\dfrac{1}{x\sqrt{x^2 - 1}}$

Chapter 9: Test yourself

1 $5e^{5x}$

2 $\frac{1}{2}e^{\frac{1}{2}x}$

3 $\dfrac{1}{2\sqrt{x}}e^{\sqrt{x}}$

4 $-2e^{-2x}$

5 $-\frac{5}{2}e^{-\frac{5}{2}x}$

6 $-2e^{(5-2x)}$

7 $ae^{(ax+b)}$

8 $2xe^{x^2}$

9 $-pe^{-px}$

10 $\dfrac{1}{a}e^{\frac{x}{a}}$

11 $\dfrac{e^x - e^{-x}}{2}$

12 $\dfrac{e^x + e^{-x}}{2}$

13 $xe^x + e^x$

14 $e^{-x} - xe^{-x}$

15 $-x^2 e^{-x} + 2xe^{-x}$

16 $xe^x + 5e^x$

17 $e^x \sin x + e^x \cos x$

18 $10e^x$

19 $2^x \ln 2$

20 $10^{2x} \times 2 \ln 10$

21 $nx^{n-1}a^x + x^n a^x \ln a$

22 $\cos x e^{\sin x}$

23 $(a+b)^x \ln(a+b)$

24 $2a^{2x+1} \ln a$

25 $2bxa^{bx^2} \ln a$

26 $-\sin x e^{\cos x}$

27 $\sec^2 x e^{\tan x}$

28 $\dfrac{1}{x}$

29 $\dfrac{2ax + b}{ax^2 + bx + c}$

30 $\dfrac{2}{x}$

31 $\dfrac{3x^2}{x^3 + 3}$

32 $1 + \ln x$

33 $\dfrac{p}{px + q}$

34 $\cot x$

35 $-\tan x$

36 $\dfrac{2a}{a^2 - x^2}$

37 $\dfrac{e^x - e^{-x}}{e^x + e^{-x}}$

38 $\dfrac{1}{\sqrt{x^2 + 1}}$

39 $\dfrac{1}{2\left(1 + \sqrt{x}\right)}$

40 $\dfrac{1}{\sin x}$

41 $\dfrac{x}{x^2 + 1}$

42 $\dfrac{e^x(2x - 1)}{2x^{\frac{3}{2}}}$

43 $2xe^{4x}(1 + 2x)$

44 $-kae^{-kx}(\sin kx - \cos kx)$

45 $\frac{1}{2}\cot x$

46 $x^x(1 + \ln x)$

47 $\dfrac{1}{2\sqrt{x^2-1}}$

48 $\dfrac{1}{1+e^x}$

49 $\dfrac{-a}{x\sqrt{a^2-x^2}}$

50 $\dfrac{2}{e^x+e^{-x}}$

Chapter 10: Test yourself

1 $\frac{1}{2}\cosh\frac{1}{2}x$

2 $2\cosh 2x$

3 $\frac{1}{3}\sinh\frac{1}{3}x$

4 $a\operatorname{sech}^2 ax$

5 $\frac{1}{4}\operatorname{sech}^2\frac{1}{4}x$

6 ae^{ax}

7 $-\dfrac{1}{x^2}\cosh\dfrac{1}{x}$

8 $\sinh 2x$

9 $3\cosh^2 x \sinh x$

10 $a\cosh(ax+b)$

11 $4x\sinh 2x^2$

12 $na\sinh^{n-1} ax\cosh ax$

13 $\cosh 2x$

14 $2\sinh 2x$

15 $2\tanh x\operatorname{sech}^2 x$

16 $2\operatorname{cosech} 2x$

17 $x\cosh x$

18 $\tanh x$

19 $3x^3\cosh 3x + 3x^2\sinh 3x$

20 1

21 $\cosh x e^{\sinh x}$

22 $\dfrac{\cosh x}{2\sqrt{\sinh x}}$

23 2

24 $\operatorname{sech}^2 x e^{\tanh x}$

25 $\dfrac{1}{\sqrt{x^2+4}}$

26 $\dfrac{1}{\sqrt{x^2-25}}$

27 $\dfrac{-2}{(1+x)\sqrt{2(1+x^2)}}$

28 $\sec x$

29 $\operatorname{sech} x$

30 $\sec x$

31 $\operatorname{sech} x$

32 $\sec x$

33 $\dfrac{2}{1-x^2}$

34 $\dfrac{-1}{x(x+2)}$

35 $\dfrac{2}{\sqrt{1+x^2}}$

36 $\dfrac{2}{\sqrt{2x(2x+1)}}$

37 $\frac{1}{2}\sec\left(\frac{x}{2}\right)$

38 $\frac{1}{2}\operatorname{sech}\left(\frac{x}{2}\right)$

39 $\ln\left(\dfrac{x+\sqrt{x^2+4}}{2}\right)$

40 $\ln\left(\dfrac{x+\sqrt{x^2-9}}{3}\right)$

41 $\ln\left(\dfrac{3x+\sqrt{9x^2-4}}{2}\right)$

42 $\dfrac{1}{2}\ln\left(\dfrac{4+x}{4-x}\right)$

Chapter 11: Exercise 11.1

1 $\frac{3}{2}x^2+C$

2 $\frac{5}{3}x^3+C$

3 $\frac{1}{8}x^4+C$

4 $0.08x^5+C$

5 $\frac{4}{3}x^9+C$

6 $5t^3+C$

7 $\frac{1}{2}x+C$

8 $\theta+C$

9 $\frac{4}{3}x^3-\frac{5}{2}x^2+x+C$

10 $\frac{3}{5}x^5-\frac{5}{4}x^4+C$

11 $\frac{8}{3}x^3-\frac{1}{4}x^2+C$

12 $\frac{6}{5}x^5+\frac{3}{2}x^4+C$

13 $\frac{1}{3}x^3-9x+C$

14 $\frac{2}{3}x^3+\frac{5}{2}x^2-12x+C$

15 $-\dfrac{1}{x}+C$

16 $-\dfrac{1}{0.4x^{0.4}}+C$

17 $-\dfrac{1}{x^3}+C$

18 $\frac{3}{4}x^{\frac{4}{3}}+C$

19 $\sqrt{x}+C$

20 $\frac{3}{2}x^{\frac{2}{3}}+C$

21 $\frac{2}{3}x^{\frac{3}{2}}+2x^{\frac{1}{2}}+C$

22 $\frac{3}{5}x^{\frac{5}{3}}+x+3x^{\frac{1}{3}}+C$

23 $\dfrac{-1}{\sqrt{2x}}+C$

24 $\frac{1}{2}\pi x-10x^{0.5}+C$

25 $gt + C$

26 $-\dfrac{1}{2x^2} + \dfrac{1}{x} + \ln|x| - x + C$

27 $\frac{2}{3}t^{\frac{3}{2}} + C$

28 $2x - \frac{1}{9}x^3 - x^{\frac{1}{2}} + C$

29 $1.4\ln|x| + C$

30 $\ln|x + 3| + C$

31 $\dfrac{1}{a}\ln|ax + b| + C$

32 $\ln\left|\dfrac{(x-1)^3}{(x-2)^4}\right| + C$

33 $\ln(x^2 + 4) + C$

34 $-\frac{1}{2}\ln|3 - 2x| + C$

35 $x + 3\ln|x| + C$

36 $\frac{1}{3}x^3 - 7\ln|x| + C$

37 $\ln|x| + \dfrac{1}{x} - \dfrac{1}{2x^2} + C$

38 $\dfrac{2}{3a}(ax + b)^{\frac{3}{2}} + C$

39 $\frac{1}{3}(2x + 3)^{\frac{3}{2}} + C$

40 $\dfrac{4}{3}\left(1 + \dfrac{x}{2}\right)^{\frac{3}{2}} + C$

41 $\dfrac{2}{a}\sqrt{ax + b} + C$

42 $-2\sqrt{1 - x} + C$

43 $\dfrac{1}{3a}(ax + b)^3 + C$

44 $\dfrac{x^2}{2} + \dfrac{x^3}{3} + \dfrac{x^4}{4} + \dfrac{x^5}{5} + C$

45 $\frac{1}{2}\ln|(x^2 - 1)| + C$

46 $-\frac{1}{a}\ln|1 + \cos ax| + C$

47 $\frac{1}{3}\ln|e^{3x} + 6| + C$

48 $\frac{1}{2}\ln|2x + \sin 2x| + C$

49 $\frac{x^4}{4} + C$

50 $2x^3 + 3$

51 $\frac{5x^3}{6} + 2x - \frac{11}{6}$

52 $y = 2x^2 - 5x + 6$

53 $y = 3x^3 - 5x^2 + 4x + 4$

54 $s = \frac{4t^3}{3} + 8t + 10$

Exercise 11.2

1 $\frac{3}{2}e^{2x} + C$

2 $\frac{1}{3}e^{3x-1} + C$

3 $\frac{1}{2}\left(e^{2x} - e^{-2x}\right) + 2x + C$

4 $ae^{x/a} + C$

5 $2\left(e^{\frac{1}{2}x} - e^{-\frac{1}{2}x}\right) + C$

6 $\frac{1}{a}\left(e^{ax} + e^{-ax}\right) + C$

7 $\frac{1}{3}\left(e^{3x} + \frac{1}{\ln a}a^{3x}\right) + C$

8 $\frac{1}{\ln 2}2^x + C$

9 $\frac{1}{3\ln 10}10^{3x} + C$

10 $\frac{1}{\ln a}(a^x - a^{-x}) + C$

11 $\frac{1}{2}e^{x^2} + C$

12 $-e^{\cos x} + C$

13 $-\frac{1}{3}\cos 3x + C$

14 $\frac{1}{5}\sin 5x + C$

15 $-2\cos\frac{1}{2}\left(x + \frac{1}{3}\pi\right) + C$

16 $\frac{1}{2}\sin(2x + a) + C$

17 $-3\cos\frac{1}{3}x + C$

18 $\frac{1}{3}\cos(a - 3x) + C$

19 $\frac{1}{a}\sin ax - \frac{1}{b}\cos bx + C$

20 $-\frac{1}{2a}\cos 2ax + C$

21 $\frac{1}{3}\sin 3x + 3\cos\frac{1}{3}x + C$

22 $\ln|x + \sin x| + C$

23 $\frac{1}{4}\sin^4 x + C$

24 $e^{\tan x} + C$

25 $\frac{1}{a}\ln|\sec ax| + \frac{1}{b}\ln|\sin bx| + C$

26 $= \ln|1 + \sin^2 x| + C$

27 $\frac{1}{2}\sinh 2x + C$

28 $\frac{3}{a}\cosh\frac{1}{3}ax + C$

29 $\frac{1}{3}\ln\cosh 3x + C$

30 $-\frac{1}{a}\cos(ax + b) + C$

31 $\frac{2}{3}e^{\frac{3}{2}x} + 4e^{\frac{1}{2}x} - 2e^{-\frac{1}{2}x} + C$

32 $\frac{2}{3}\ln\left|\sec\frac{3}{2}x\right| + C$

33 $3\tan\frac{1}{3}x + C$

34 $\ln|1 + e^x| + C$

35 $\ln|1 + \tan x| + C$

36 $\frac{2}{3}(\sin x)^{\frac{3}{2}} + C$

Exercise 11.3

In some of these answers there is an alternative logarithmic form.

1 $\sin^{-1}\frac{x}{3} + C$

2 $\cosh^{-1}\frac{x}{3} + C$

3 $\sinh^{-1}\frac{x}{3} + C$

4 $\frac{1}{3}\tan^{-1}\frac{x}{3} + C$

5 $\frac{1}{3}\tanh^{-1}\frac{x}{3} + C$

6 $-\frac{1}{3}\coth^{-1}\frac{x}{3} + C$

7 $\sin^{-1}\dfrac{x}{4} + C$

8 $\dfrac{1}{4}\tanh^{-1}\dfrac{x}{4} + C$

9 $\cosh^{-1}\dfrac{x}{4} + C$

10 $-\dfrac{1}{4}\coth^{-1}\dfrac{x}{4} + C$

11 $\sinh^{-1}\dfrac{x}{4} + C$

12 $\dfrac{1}{4}\tan^{-1}\dfrac{x}{4} + C$

13 $\dfrac{1}{3}\sin^{-1}\dfrac{3x}{5} + C$

14 $\dfrac{1}{3}\cosh^{-1}\dfrac{3x}{5} + C$

15 $\dfrac{1}{3}\sinh^{-1}\dfrac{3x}{5} + C$

16 $\dfrac{1}{6}\tan^{-1}\dfrac{2x}{3} + C$

17 $\dfrac{1}{6}\tanh^{-1}\dfrac{2x}{3} + C$

18 $-\dfrac{1}{6}\coth^{-1}\dfrac{2x}{3} + C$

19 $\dfrac{1}{6}\tan^{-1}\dfrac{3x}{2} + C$

20 $\dfrac{1}{3}\sinh^{-1}\dfrac{3x}{2} + C$

21 $\dfrac{1}{3}\cosh^{-1}\dfrac{3x}{2} + C$

22 $\dfrac{1}{7}\sinh^{-1}\dfrac{7x}{5} + C$

23 $\dfrac{1}{\sqrt{2}}\sinh^{-1}\left(x\sqrt{\dfrac{2}{5}}\right) + C$

24 $\sin^{-1}\dfrac{x}{\sqrt{5}} + C$

25 $\dfrac{1}{2}\sin^{-1}\dfrac{2x}{\sqrt{5}} + C$

26 $\dfrac{1}{2}\sinh^{-1}\left(\dfrac{2x}{\sqrt{5}}\right) + C$

27 $\dfrac{1}{10}\tanh^{-1}\dfrac{2x}{5} + C$

28 $\dfrac{1}{\sqrt{7}}\sinh^{-1}\left(\dfrac{x\sqrt{7}}{6}\right) + C$

29 $-\cosech^{-1} x + C$

30 $\dfrac{1}{2}\tan^{-1}\left(\dfrac{\sqrt{x^2-4}}{2}\right) + C$

31 $-\dfrac{1}{2}\cosech^{-1}\dfrac{x}{2} + C$

32 $-\dfrac{1}{2}\sech^{-1}\dfrac{x}{2} + C$

Chapter 12: Exercise 12.1

1 $\frac{1}{2}(x - \sin x) + C$

2 $\frac{1}{2}(x + \sin x) + C$

3 $2\tan\frac{1}{2}x - x + C$

4 $\frac{3}{8}x + \frac{1}{4}\sin 2x + \frac{1}{32}\sin 4x + C$

5 $\frac{3}{8}x - \frac{1}{4}\sin 2x + \frac{1}{32}\sin 4x + C$ **6** $-\frac{1}{2}\cot 2x + x + C$

7 $\frac{1}{2}x - \frac{1}{8}\sin 4x + C$ **8** $\frac{1}{2}x + \frac{1}{12}\sin 6x + C$

9 $\frac{1}{2}x + \frac{1}{4a}\sin 2(ax + b) + C$ **10** $-\frac{3}{4}\cos x + \frac{1}{12}\cos 3x + C$

11 $\frac{1}{12}\sin 3x + \frac{3}{4}\sin x + C$ **12** $\frac{1}{2}\sin x - \frac{1}{10}\sin 5x + C$

13 $\frac{1}{4}\sin 2x + \frac{1}{8}\sin 4x + C$ **14** $-\frac{1}{4}\cos 2x - \frac{1}{12}\cos 6x + C$

15 $-\frac{1}{11}\cos\frac{11}{2}x - \frac{1}{5}\cos\frac{5}{2}x + C$ **16** $\frac{1}{8}x - \frac{1}{32}\sin 4x + C$

17 $\tan x - \cot x + C$ **18** $2\tan x - x + C$

19 $\frac{1}{2}\tan^2 x - \ln|\sec x| + C$ **20** $2\sqrt{2}\sin\frac{1}{2}x + C$

Exercise 12.2

1 $\frac{25}{2}\sin^{-1}\frac{1}{5}x + \frac{1}{2}x\sqrt{25 - x^2} + C$

2 $\frac{9}{2}\sin^{-1}\frac{1}{3}x + \frac{1}{2}x\sqrt{9 - x^2} + C$

3 $\frac{1}{4}\sin^{-1} 2x + \frac{1}{2}x\sqrt{1 - 4x^2} + C$

4 $\frac{9}{4}\sin^{-1}\frac{2}{3}x + \frac{1}{2}x\sqrt{9 - 4x^2} + C$

5 $-\dfrac{\sqrt{1 - x^2}}{x} + C$ **6** $-\dfrac{\sqrt{a^2 + x^2}}{a^2 x} + C$

7 $2\ln\left|\tan\frac{1}{4}x\right| + C$ **8** $2\ln\left|\tan\left(\frac{1}{4}x + \frac{1}{4}\pi\right)\right| + C$

9 $\frac{1}{3}\ln\left|\tan\frac{3}{2}x\right| + C$ **10** $\ln|\tan x| + C$

11 $\tan\frac{1}{2}x + C$ **12** $\tan x - \sec x + C$

13 $\tan x + \sec x + C$ **14** $\ln\left|\dfrac{1}{1 - \sin x}\right| + C$

15 $\frac{1}{2}\tan^{-1}\left(\frac{1}{2}\tan\frac{1}{2}x\right) + C$ **16** $\frac{1}{2}\tan^{-1}\left(2\tan\frac{1}{2}x\right) + C$

Exercise 12.3

1 $\frac{1}{3}\sin x^3 + C$

2 $\frac{1}{6}\ln\left|\dfrac{1}{1-2x^3}\right| + C$

3 $\sqrt{1+x^2} + C$

4 $-\frac{2}{5}\sqrt{2-5x} + C$

5 $-2\cos\sqrt{x} + C$

6 $\frac{2}{3}\sqrt{1+x^3} + C$

7 $\frac{1}{2}\ln\left|\dfrac{1}{1+2\cos x}\right| + C$

8 $\frac{1}{2}\left(\ln|x|\right)^2 + C$

9 $\frac{1}{3}(5+x^2)^{\frac{3}{2}} + C$

10 $\tan^{-1}x^2 + C$

11 $\frac{1}{30}(x-2)^5(5x+2) + C$

12 $\ln|x+1| + \dfrac{4x+3}{2(x+1)^2} + C$

13 $\frac{2}{3}(x+2)\sqrt{x-1} + C$

14 $\frac{2}{15}(x-1)^{\frac{3}{2}}(3x+2) + C$

15 $-\sqrt{5-x^2} + C$

16 $\frac{1}{3}(x^2+2)\sqrt{x^2-1} + C$

17 $\frac{1}{15}(x^2-2)^{\frac{3}{2}}(3x^2+4) + C$

18 $2\sqrt{x} + 6\ln\left|\sqrt{x}-3\right| + C$

19 $\frac{1}{5}\cos^5 x - \frac{1}{3}\cos^3 x + C$

20 $\frac{1}{3}\sin^3 x - \frac{2}{5}\sin^5 x + \frac{1}{7}\sin^7 x + C$

Exercise 12.4

1 $\sin x - x\cos x + C$

2 $\frac{1}{9}\sin 3x - \frac{1}{3}x\cos 3x + C$

3 $(x^2-2)\sin x + 2x\cos x + C$

4 $\frac{1}{2}x^2\ln|x| - \frac{1}{4}x^2 + C$

5 $e^x(x-1) + C$

6 $e^x(x^2-2x+2) + C$

7 $e^{-ax}\left(\dfrac{-ax-1}{a^2}\right) + C$

8 $\frac{1}{5}e^x(\cos 2x + 2\sin 2x) + C$

9 $x\cos^{-1}x - \sqrt{1-x^2} + C$

10 $x\tan^{-1}x - \frac{1}{2}\ln(1+x^2) + C$

11 $\frac{1}{2}(x^2+1)\tan^{-1}x - \frac{1}{2}x + C$

12 $\frac{1}{2}e^x(\sin x - \cos x) + C$

13 $-\frac{1}{4}x\cos 2x + \frac{1}{8}\sin 2x + C$

14 $x\tan x - \ln|\sec x| + C$

Chapter 13: Exercise 13.1

1 $x - 2\ln|x + 2| + C$

2 $-x - \ln|1 - x| + C$

3 $\dfrac{1}{b^2}\left(bx - a\ln|a + bx|\right) + C$

4 $x + 2\ln|x - 1| + C$

5 $-x + 2\ln|x + 1| + C$

6 $x - 2\ln|2x + 3| + C$

7 $\frac{1}{2}x^2 - 2x + 4\ln|x + 2| + C$

8 $-\frac{1}{2}x^2 - x - \ln|1 - x| + C$

Exercise 13.2

1 $\dfrac{1}{2}\ln\left|\dfrac{x - 1}{x + 1}\right| + C$

2 $\dfrac{1}{2}\ln\left|\dfrac{1 + x}{1 - x}\right| + C$

3 $x + \ln\left|\dfrac{x - 2}{x + 2}\right| + C$

4 $\dfrac{1}{12}\ln\left|\dfrac{2x - 3}{2x + 3}\right| + C$

5 $3\ln|x + 2| - 2\ln|x + 4| + C$

6 $2\ln|x + 3| + \ln|x - 2| + C$

7 $\ln|2x + 5| + 3\ln|x - 7| + C$

8 $\frac{2}{5}\ln|x - 1| - \frac{1}{15}\ln|3x + 2| + C$

9 $3\ln|x + 1| - \frac{5}{4}\ln|4x - 1| + C$

10 $\ln|1 - x| + \dfrac{2}{1 - x} + C$

11 $2\ln|x + 2| + \dfrac{5}{x + 2} + C$

12 $\dfrac{1}{2}\ln|2x + 3| + \dfrac{1}{2x + 3} + C$

13 $x + 2\ln|x - 4| - \ln|x + 3| + C$

14 $x + \frac{5}{3}\ln|x - 2| - \frac{2}{3}\ln|x + 1| + C$

Exercise 13.3

1 $\dfrac{1}{\sqrt{8}}\tan^{-1}\dfrac{x + 3}{\sqrt{8}} + C$

2 $\dfrac{1}{2\sqrt{13}}\ln\left|\dfrac{x + 3 - \sqrt{13}}{x + 3 + \sqrt{13}}\right| + C$

3 $\dfrac{1}{\sqrt{2}}\tan^{-1}\dfrac{x + 2}{\sqrt{2}} + C$

4 $\dfrac{1}{\sqrt{13}}\tan^{-1}\dfrac{2x + 1}{\sqrt{13}} + C$

5 $-\dfrac{1}{2}\ln|3x^2 + 4x + 2| + \dfrac{3}{\sqrt{2}}\tan^{-1}\dfrac{3x + 2}{\sqrt{2}} + C$

6 $2\ln|x^2 - 2x - 1| - \dfrac{\sqrt{2}}{4}\ln\left|\dfrac{x - 1 - \sqrt{2}}{x - 1 + \sqrt{2}}\right| + C$

7 $\ln|x^2 + 4x + 5| + \tan^{-1}(x + 2) + C$

8 $\dfrac{1}{3}\ln|x + 1| - \dfrac{1}{6}\ln|x^2 - x + 1| + \dfrac{1}{\sqrt{3}}\tan^{-1}\dfrac{2x - 1}{\sqrt{3}} + C$

9 $x - 2\ln|x^2 + 2x + 2| + 3\tan^{-1}(x + 1) + C$

10 $-\dfrac{3}{2}\ln\left|x + \sqrt{2} + 1\right| + \dfrac{1}{\sqrt{2}}\ln\left|x + \sqrt{2} - 1\right| + C$

  $-\dfrac{3}{2}\ln\left|x^2 + 2x + 1\right| - \dfrac{1}{\sqrt{2}}\ln\left|\dfrac{x + 1 + \sqrt{2}}{x + 1 - \sqrt{2}}\right| + C$

Exercise 13.4

1 $\ln\left|x + 3 + \sqrt{x^2 + 6x + 10}\right| + C$ **2** $\ln\left|x + 1 + \sqrt{x^2 + 2x + 4}\right| + C$

3 $\ln\left|x - 2 + \sqrt{x^2 - 4x + 2}\right| + C$ **4** $\sin^{-1}\dfrac{2x + 1}{\sqrt{5}} + C$

5 $\sin^{-1}\dfrac{x - 2}{2} + C$ **6** $\sqrt{x^2 + 1} + C$

7 $\sqrt{x^2 + 1} + \ln\left|x + \sqrt{x^2 + 1}\right| + C$ **8** $\sqrt{x^2 - 1} + \ln\left|x + \sqrt{x^2 - 1}\right| + C$

9 $\sqrt{x^2 - x + 1} + \dfrac{1}{2}\ln\left|2x - 1 + \sqrt{x^2 - x + 1}\right| + C$

10 $2\sqrt{x^2 - x + 5} - \ln\left|x - 1 + \sqrt{x^2 - x + 5}\right| + C$

11 $-3\sin^{-1}\dfrac{x + 2}{\sqrt{7}} - 2\sqrt{3 - 4x - x^2} + C$

12 $\sqrt{x^2 + 2x - 1} + \ln\left|x + 1 + \sqrt{x^2 + 2x - 1}\right| + C$

Chapter 14: Exercise 14.1

1 21 **2** $4\frac{1}{3}$

3 $1\frac{5}{6}$ **4** 9

5 $5 \times 10^{\frac{1}{3}} - 5$ **6** $4\frac{2}{3}$

7 $9\frac{1}{3}$ **8** $\frac{1}{3}$

9 0 **10** 0

11 $\frac{3}{2}\ln 2$ 12 $2(e-1)$

13 $\frac{1}{4}\pi r^2$ 14 $8\pi\rho a^5/15$

15 $(e^{kb}-e^{ka})/k$ 16 $\frac{1}{4}\pi$

17 $\ln\sqrt{2}$ 18 1

19 $-\frac{1}{4}$ 20 $-\frac{1}{9}$

Exercise 14.2

If no answer is given the integral does not exist.

2 $\frac{1}{8}$ 3 $\frac{1}{2}\pi$

4 $\frac{1}{2}\ln 3$ 6 1

7 $\frac{1}{2}$ 9 $1-\ln 2$

10 $\ln 2 - \frac{1}{2}$ 12 $\frac{1}{2}\pi$

13 π 14 $-\frac{1}{4}$

16 2   17 -1

18 2   20 0

Chapter 15: Exercise 15.1

1 $152\frac{1}{4}$ 2 $36\frac{3}{4}$

3 $5\ln 5 - 4$ 4 $2\frac{1}{3}$

5 $25\frac{1}{3}$ (or $50\frac{2}{3}$ for area below also)   6 4π

7 6.200 and 3.626   8 12π

9 2.283   10 $4\ln 2$

11 $\frac{4}{27}$ 12 $e^3 - 1$

13 $4\frac{1}{2}$ 14 $25\frac{3}{5}$

15 $1\frac{1}{2} - 2\ln 2$ 16 $21\frac{1}{12}$

17 $341\frac{1}{3}$ 18 $\frac{1}{3}$

19 $e - e^{-1} = 2.350$ 20 40

21 $\frac{1}{32}$

Exercise 15.2

1 $\frac{3}{2}\pi a^2$

2 $\frac{1}{8}\pi a^2$, 4

3 $\frac{1}{2}a^2$, 2

4 $\frac{4}{3}\pi^3 a^2$

5 $\frac{4}{3}a^2\left(\text{Use } \sec^4\frac{1}{2}\theta = \sec^2\frac{1}{2}\theta\left(1+\tan^2\frac{1}{2}\theta\right)\right)$

6 $29\frac{1}{2}\pi$

Exercise 15.3

1 $\dfrac{2}{\pi}$

2 0

3 $\dfrac{\ln 10}{9}$

4 $2\frac{2}{3}$

5 $\dfrac{b}{2}$

6 $\dfrac{2\sqrt{2}}{\pi}$

7 $\dfrac{2a}{\pi}$

8 $\dfrac{2v_0^2}{\pi g}$

Chapter 16: Test yourself

1 259 m²

2 6.24 m²

3 73.5 m²

4 60.7 mm²

5 1370 m²

7 1.11700

8 1.11137 (a more accurate result is 1.11145)

Chapter 17: Test yourself

1 a $\frac{243}{5}\pi$ b 8π

2 a $\frac{2187}{7}\pi$ b $\frac{96}{5}\pi$

3 $\frac{1}{6}\pi$

4 a $\frac{20}{3}\pi$ b $\frac{8}{3}\pi$

5 $\frac{64}{3}\pi$

6 32π

7 $\frac{4}{3}\pi a^3$

8 $\frac{384}{7}\pi$

9 $\frac{1}{2}\pi^2$

10 $\frac{16}{15}\pi$

11 $\frac{3}{4}\pi$

12 $\frac{96}{5}\pi$

Chapter 18: Test yourself

1 $\sqrt{5} + \frac{1}{2}\ln(2 + \sqrt{5})$

2 $2\sqrt{5} + \ln(2 + \sqrt{5})$

3 $12\frac{11}{27}$

4 $\frac{1}{2}\left(e - \frac{1}{e}\right)$

5 $\sqrt{5} - \sqrt{2} + \ln\frac{\sqrt{5} - 1}{2(\sqrt{2} - 1)}$

6 $\ln(2 + \sqrt{3})$

7 $2\pi a$

8 $\frac{3}{2}\pi a$

Chapter 19: Test yourself

1 $\sin a + x\cos a - \frac{x^2}{2!}\sin a$

2 $\cos a - x\sin a + \frac{x^2}{2!}\cos a$

3 $e^b\left(1 + x + \frac{x^2}{2!}\right)$

4 $x - \frac{1}{2}x^2 + \frac{1}{6}x^3$

5 $1 - \frac{x^2}{2!} + \frac{x^2}{4!}$

6 $x + \frac{1}{3}x^3 + \frac{2}{15}x^5$

7 $\ln 2 + \frac{1}{2}x + \frac{1}{8}x^2$

8 $1 + x\ln a + \frac{x^2(\ln a)^2}{2!}$

9 $1 - kx + \frac{k^2x^2}{2!}$

10 $1 + x + \frac{1}{2}x^2$

11 $1 + \frac{x^2}{2!} + \frac{5x^4}{4!}$

12 $\frac{1}{2}x^2 + \frac{1}{12}x^4 + \frac{1}{45}x^6$

13 $x + \frac{1}{6}x^3 + \frac{3}{40}x^5$

14 $-x - \frac{1}{2}x^2 - \frac{1}{3}x^3$

15 $x + \frac{x^3}{3!} + \frac{x^5}{5!}$

16 $x + x^2 + \frac{1}{3}x^3$

17 $x - \frac{1}{3}x^3 + \frac{2}{15}x^5$

Chapter 20: Exercise 20.1

1 $y = \frac{k}{x} + c$

2 $y = ce^{\frac{x}{a}}$

3 $y = cx$

4 $(1 + y)(1 - x) = c$

5 $\frac{y}{x + 1} = c$

6 $\sec x = c \sec y$ or $\cos x = k \cos y$

7 $(1 + y^2)(1 + x^2) = cx^2$

8 $\ln|x^2y| - y = c$

9 $\dfrac{1+y}{1-y} = c\tan x$

10 $y = ce^{\frac{1}{3}x^3}$

11 $\sqrt{x^2 - 1} - \sqrt{y^2 - 1} = c$

12 $\frac{1}{2}x^2 + \ln x - \frac{1}{3}y^3 - \frac{1}{2}y^2 = c$

13 $y = cex^2$

14 $xy = c$

Exercise 20.2

1 $y + 1 = ce^{x^2}$

2 $x^2 + 2xy = c$

3 $y = x + 1 + ce^x$

4 $y = ce^{-\frac{1}{2}x^2} + 1$

5 $y = \dfrac{e^x}{a+1} + ce^{-ax}$

6 $y \sec x = \ln(\sec x + \tan x) + c$

7 $y = cx^a + \dfrac{x}{1-a} - \dfrac{1}{a}$

8 $y + 1 = c \sin x$

9 $ye^x = x + c$

10 $y = \dfrac{x}{1-a} - \dfrac{1}{a} + cx^a$

11 $y = (\tan x - 1) + ce^{-\tan x}$

12 $xy + \ln|x| = c$

Exercise 20.3

1 $y = Ae^{3x}$

2 $y = Ae^{-4x}$

3 $y = Ae^{2x} + Be^{4x}$

4 $y = Ae^{2x} + Be^{-7x}$

5 $y = A + Be^{-5x}$

6 $y = Ae^{2x} + Be^{-2x}$

7 $y = A \sin 2x + B \cos 2x$

8 $y = e^{-x}(A \sin 2x + B \cos 2x)$

9 $y = Ax + B$

10 $y = e^{-3x}(Ax + B)$

11 $y = -1 + Ae^{3x}$

12 $y = x + 1 + Ae^{-4x}$

13 $y = Ae^{2x} + Be^{4x} + \cos x$

14 $y = -\frac{1}{8}e^x + Ae^{2x} + Be^{-7x}$

15 $y = x + A + Be^{-5x}$

16 $y = \frac{1}{4}xe^{2x} + Ae^{2x} + Be^{-2x}$

17 $y = \frac{1}{8}e^{2x} + A\sin 2x + B\cos 2x$

18 $y = 2 + e^{-x}(A \sin 2x + B \cos 2x)$

19 $y = \frac{1}{3}x^3 + \frac{1}{2}x^2 + Ax + B$

20 $y = \frac{1}{36}e^{3x} + e^{-3x}(Ax + B)$

Test yourself

1 $x^2 + 2xy = c$ 2 $y = x(\ln x + c)$

3 $\ln\sqrt{x^2 + y^2} - \tan^{-1}\dfrac{y}{x} = c$ 4 $\dfrac{x}{x - y} = \ln\left|\dfrac{y - x}{c}\right|$

5 $x^2 - y^2 = cx$ 6 $\dfrac{1}{x} - \dfrac{1}{y} = c$

7 $xy(x - y) = c$ 8 $xy^2 = c(x + 2y)$

9 $y = ce^{\frac{y}{x}}$ 10 $xy^2 = c(x + 2y)$

11 $xy + \ln|x| = c$

Chapter 21: Test yourself

1 800 metres

2 9.58 grams

3 About 115 minutes

Index